ROUTLEDGE LIBRARY EDITIONS:
ECONOMIC GEOGRAPHY

T0227528

URBANISM, COLONIALISM, AND THE WORLD-ECONOMY

URBANISM, COLONIALISM, AND THE WORLD-ECONOMY

Cultural and Spatial Foundations of the World Urban System

ANTHONY D. KING

Routledge
Taylor & Francis Group

LONDON AND NEW YORK

First published in 1990

This edition first published in 2015
by Routledge
2 Park Square, Milton Park, Abingdon, Oxon, OX14 4RN

and by Routledge
711 Third Avenue, New York, NY 10017

Routledge is an imprint of the Taylor & Francis Group, an informa business

British Library Cataloguing in Publication Data
A catalogue record for this book is available from the British Library

ISBN: 978-1-138-85764-3 (Set)
eISBN: 978-1-315-71580-3 (Set)
ISBN: 978-1-138-88533-2 (Volume 4)
eISBN: 978-1-315-71550-6 (Volume 4)
Pb ISBN: 978-1-138-88534-9 (Volume 4)

Publisher's Note
The publisher has gone to great lengths to ensure the quality of this reprint but points out that some imperfections in the original copies may be apparent.

Disclaimer
The publisher has made every effort to trace copyright holders and would welcome correspondence from those they have been unable to trace.

URBANISM, COLONIALISM, AND THE WORLD-ECONOMY

Cultural and Spatial Foundations
of the
World Urban System

ANTHONY D. KING

London and New York

First published 1990
Reprinted in paperback 1991
by Routledge
11 New Fetter Lane, London EC4P 4EE
29 West 35th Street, New York, NY 10001

Phototypeset in 10pt Baskerville by
Mews Photosetting, Beckenham, Kent
Printed and bound in Great Britain by
Biddles Ltd, Guildford and King's Lynn

British Library Cataloguing in Publication Data

King, Anthony D. (Anthony Douglas), *1931–*
Urbanism, colonialism and the world-
economy: cultural and spatial foundations
of the world urban system – (International library of sociology)
1. Urban regions. Social aspects
I. Title II. Series
307.7'6

Library of Congress Cataloging-in-Publication Data
King, Anthony D.
Urbanism, colonialism, and the world-economy: cultural and spatial
foundations of the world urban system /
Anthony D. King.
p.   cm. – (International library of sociology)
Bibliography: p.
Includes indexes.
1. Urbanization — Economic aspects. 2. Urbanization — Social
aspects — Great Britain — colonies. 3. Architecture and society —
colonies. 4. Sociology, Urban. I. Title. II. Series.
HT361.K567   1989
307.7'6—dc 20   88-13827
                                                            CIP
ISBN 0-415-06240-3

CONTENTS

TABLES

FIGURES

PREFACE

Over the last few years, there has been an explosion of consciousness concerning the impact of world-economic forces on cities. In Britain, this has been given greater momentum by the economic and social effects of City deregulation ('Big Bang' of October 1986), enabling it to compete more freely in world financial markets and at the same time, exacerbating the economic and social divide between the patchy prosperity of 'financial services London' and the deindustrialized cities of the North.

In the United States, too, as well as continental Europe, changes in cities are increasingly seen within a larger global context. The 'export of jobs', the quest for investment, and the internationalization of the economy generally have turned people's attention to looking at the world beyond their own state.

Both the causes for as well as the solutions to these urban problems are sought in the larger world-economy. Academics examine aspects of global restructuring while cities look to global markets to revive their fading fortunes. At the time and place of writing (November, 1987, Yorkshire), Sheffield is celebrating its success in securing the World Student Games in 1991, hoping this will provide a future economic base in sport in what was once Britain's premier 'steel city'. Leeds has recently entered the world entertainment circuit hosting (at an outdoor stadium within earshot of where this is being written) the first European concert of American pop star Madonna's world tour. In both cities, these examples of the 'mobilization of the spectacle' are typical of the new post-modern economy (Harvey, 1987).

At one time a world centre for the manufacture of ready-made clothes, Leeds is now — with the financial services' restructuring of banking, building societies, and the real-estate market — being referred to as the 'financial centre of the North'. According to the Yorkshire and Humberside Development Association's *Development Review* (1987) (also registered and distributed in Palo Alto, California), there were almost 300 branches of United States corporations in Yorkshire and Humberside, more than half of them having manufacturing interests (1 March 1987, p. 22). And in what was, in 1980, an almost totally publicly funded system of higher education in Britain, some urban universities now earn as much as 10–15 per cent of their income from international students' fees (*Times Higher Educational Supplement*, 2 October 1987, p. 4).

This marked shift towards the further internationalization of the economy has been charted in many studies, not least of the growth of 'global cities' in the world-economy, the subject of the companion volume (King, 1990). Yet in researching that book, it became apparent that the continuities with the past are at least as great as the discontinuities. Most of the urban foundations of the present world-economy were laid in the colonial era, especially in the nineteenth and first half of the twentieth centuries. This is true not only for cities in Asia, Latin America, Africa, and the Far East but also, for countries at the core of the world-economy such as Britain. (There is, indeed, more than a touch of irony in the fact that the recent 'official regeneration' of the port of Liverpool, whose economic growth and decline were very much tied up with the growth and decline of imperial trade, was commemorated with the opening of a northern extension of another London-based institution with its origins in colonial trade, the Tate Gallery.)

In short, any understanding of contemporary urbanism requires an understanding of its colonial past and for these reasons, I have revised, redeveloped, and extended earlier work in this field as a necessary background to understanding today's 'global cities'.

However, in addition to the substantive issues that these essays deal with, they are also concerned with methodology.

Despite the moves to 'interdisciplinarity' over the years, much of academic research is generally undertaken in different 'camps', largely defined by disciplinary departments (which, in turn, define career structures) and, in a looser sense, by participation in particular conferences or professional groups. Such specialization is, of course, necessary and productive, but it can also be dysfunctional. In the belief that there is much to be gained by incorporating approaches from different research specializations, the essays in this book attempt to draw on, and address four or five of them, each represented by a cluster of professional associations and discourses. As the different approaches are discussed in the following pages, they will, at the risk of overgeneralization, only be referred to briefly here. However, it is readers in these fields to whom the book is addressed.

The first is the growing body of work in urban political economy (once called 'the new urban sociology' or sometimes, 'critical urban theory') that, until quite recently, has neglected the study of building form, architecture, and urban form either as a source of data for understanding economic and social change or as an object of study in itself. And as this (particularly in sociology) is principally concerned with contemporary developments, there have been relatively few historical studies.

The second is work in urban and, especially, planning history (which has also grown considerably in recent years) with its interest in the physical and spatial form of cities though, on the whole, less inclined towards the concerns of 'the new urban sociology'.

The third tradition is represented by a combination of approaches from different social sciences (geography, sociology, psychology, and anthropology) as well as architecture and planning, the focus of which is on 'the study of people and their physical surroundings', 'the mutual interaction of people and their built environment', 'built form and culture', 'man[*sic*]–environment studies', to use some of the terms encompassing this field. These studies, of perhaps more interest to the design professions, have again, not given rise to much work of a historical nature and, with exceptions, have remained distant from work in urban political economy (King, 1987). More recently, however, some of these approaches are being taken up in planning history (Rogers, 1987).

The fourth tradition includes studies in architecture in which historical work figures prominently, though where often the 'dominant mode of analysis', to quote Blau (1984: 91), 'is one of interpretation'.

Finally, there is an equally growing corpus of work in world-system studies that is being applied increasingly to urban and other issues, though to date it seems not to have been much used for the study of architecture, building form, and urban form. This category may also be appropriate to refer to research in 'development studies' whether these focus on countries in the core or the periphery.

None of these approaches mentions work in different spheres of geography yet it will be apparent from both the bibliography and much of what follows that the particular urban and spatial focus of geography is central to the topics discussed.

To try and bring something of these approaches together is a tall order, and probably an overambitious task. And listing them at the start of a book is providing a 'hostage to fortune' for any reviewer. None the less, it is because the understanding of contemporary political, economic, social, cultural, architectural, and urban phenomena requires something from all these perspectives that the following chapters draw from each of them and, hopefully, by so doing, bring them closer together.

Finally, I have not included in these chapters any discussion of the North American colonies (and cities) before 1776 nor of the USA after that date and their relation to the colonial system. To view the process of North American colonization in a theoretical framework that incorporates the colonial experience of South America, Australasia, Africa, and South and South-East Asia, particularly with regard to relations with the continent's indigenous inhabitants and the cities that developed, is a research task that would have very considerable potential in revealing new insights, both for urban theory and history as well as for the understanding of North American cities (and societies) themselves. Whilst the place of North America in the early colonial system is encompassed in the framework set out in Chapter 2 and the influence of the USA is also discussed extensively in the context of London's role as 'world city' in *Global Cities* (1989c), the influence of the United States on British urban development is also another issue that will need to wait for later attention.

ACKNOWLEDGEMENTS

Some of the chapters in this book started out as lectures or papers first produced at the invitation of particular organizations and I would like to thank them here: the Centre for the History of European Expansion, University of Leiden (Chapter 2); the editors of the Urban History Yearbook, University of Leicester (Chapter 4); the Aga Khan Program in Islamic Architecture, Harvard/MIT (Chapter 5), and the Department of Sociology, University of Amsterdam and Netherlands Association of Sociology and Anthropology (Chapter 6).

Mike Safier, of the Development Planning Unit, Bartlett School of Architecture and Planning, University College, London, provided critical comments and many suggestions on earlier drafts of the papers and I would like to acknowledge his valuable suggestions over my years of association with the DPU. In addition, many others have helped by sending their own papers or commenting on earlier versions of them and I would like to take this opportunity of thanking Janet Abu-Lughod, George Atkinson, John Belcher, Ray Betts, Gordon Cherry, Leon Deben, Peter Gleichmann, Peter Hall, Deryck Holdsworth, Hubert Morsink, Chris Pickvance, Paul Rabinow, Robert Reed, David Simon, Gwen Wright, and others, though not mentioned, to whom I am very grateful. At Routledge, I would like to thank Chris Rojek and John Urry (the series editor), for their help and valuable advice and also, Eve Daintith for her sympathetic and meticulous copy-editing.

As always, my thanks are also due to Ursula, Frances, Karen, Anna, Nina, and also to David; needless to say, the responsibility for what follows is entirely my own.

Chapter One

INTRODUCTION

Urbanism, colonialism, and the world-economy

In choosing the title for this book, I have three objectives in mind. First, to focus on urbanism, particularly urban forms and processes around the world, and especially in one-time colonial societies that, especially from the last century and particularly, in the last four decades, have become increasingly integrated in a single world-economy. And when compared to the variety of forms that existed before this time, they have also become increasingly similar.

'Urbanism' also emphasizes the symbiotic relationship between the material and spatial aspects of cities, their built environment, and architectural form, and the social, economic, and cultural systems of which they are a part. The book, therefore, is concerned with understanding not only what has increasingly been called the 'social production of the built environment' but also, how built environments both represent and condition economies, societies, and cultures. Like Prior (1988), I would maintain that the built environment is more than a mere representation of social order (i.e. a reflector), or simply a mere environment in which social action takes place. Rather, physical and spatial urban form actually constitute as well as represent much of social and cultural existence: society is to a very large extent constituted through the buildings and spaces that it creates. It is for these reasons that many of the following chapters focus on the built environment.

More specifically, however, the chapters are concerned with the forces that have created contemporary societies and environments, especially those operating at a global scale.

The 1980s have seen an immense growth in consciousness concerning the impact of world-economic forces on cities. Yet whilst this global *consciousness* is new, the phenomena themselves are not: for the past three centuries at least, world-economic, political, and cultural

1

forces have been major factors shaping cities, the spatial organization of societies, patterns of urbanization, and the physical and spatial forms of the built environment — but they have been insufficiently recognized and their investigation has been largely neglected. It is a central argument of this book that the historical context of contemporary global restructuring must be recognized if present-day urban and regional change, as also the class, cultural, racial, and ethnic composition of cities, are to be properly understood. Hence, my second objective is to state the case for a more historically informed approach to the growing corpus of work on the global context of urban growth.

I have also used 'world urban system' as well as 'world-economy'. This is to emphasize, perhaps more by way of metaphor, that historically, the world has increasingly become one large, interdependent city: interdependent in that, in major cities of the world (both 'East/West' as well as 'North/South'), it is not only people, knowledge, images, and ideas that move between them but also, to varying degrees, capital, labour, and goods. The world is increasingly organized through a single, interacting and interdependent urban system, even though flows between its various component parts are immensely uneven.

The chapters, therefore, explore the historical foundations of this world urban system, its spatial and cultural links, if not in its entirety (I do not, for example, deal with large parts of Asia, including China, or Russia) but rather, a major part of it — namely, that part influenced by European colonialisms. The object is not to attempt a comprehensive account, backed by a battery of empirical data, but rather, to suggest a series of approaches and methods to stimulate further research. It is, however, the attention that the chapters give to colonialism as the historical link between urbanism and the contemporary world-economy that forms the third main objective of the book.

URBANISM AND COLONIALISM

Considering its impact on contemporary urban, political, economic, social, and cultural life, the historical experience of colonialism and imperialism is greatly under-researched. Despite the fact that virtually all peripheral regions in the world-economy were at one time controlled by European core powers for varying periods between 1500 and 1950, creating 'a world organised as one huge functional region of the core states' and which remains 'the dominant spatial organisation of the twentieth century' (Taylor, 1985: 67–8), the subject of imperialism is still largely neglected. This is particularly true with regard to its significance in structuring patterns of urbanization and urbanism. The following account, therefore, first looks at imperialism and colonialism in the development of the world–economy (drawing on the account of Taylor, 1985, Chapter 3) and then focuses on the role of cities within this. As Taylor states, the formal political control of parts of the periphery has been a feature of the world-economy since its inception; formal (and subsequently informal) imperialism has been a common strategy of core domination over the periphery.

Drawing on the work of Bergeson and Schoenberg (1980), Taylor (1985) demonstrates the existence of two long waves of colonial expansion and contraction, the first, from 1500 to 1800, the second, from 1800 to 1925, related to long wave cycles of expansion and decline in the world-economy. Using the presence of a colonial governor to indicate the imposition of sovereignty of a core state over territory in the periphery, Bergeson and Schoenberg (1980) indicate the existence of 412 colonial jurisdictions between 1415 and 1969. We can look first at the imperial states in the core.

In the history of the world-economy, there have been twelve formal imperial states, only five of which have been major colonizers: Spain, Portugal, and the Netherlands (principally between 1500 and 1750); and France and Great Britain, from 1600 to 1925. Also in this first phase (especially 1600 to 1750) are the first 'minor' colonizing states of the Baltic: Denmark, Sweden, and Brandenburg/Prussia.

In the second wave, from 1870, Belgium, Germany, Italy, Japan, and the USA enter as 'late comers' (in a more detailed table on 'the seizure of territories by imperial powers, 1860–1913, Chirot (1986: 76–80) also lists the colonial activities of Russia, which took Ussuri from China in 1860, and Austria-Hungary, which took Bosnia-Herzogovina from the Ottoman Empire in 1878).

In the periphery, Taylor (1985) lists fifteen 'political arenas' in which colonial activity occurred. The first is Iberian America, including Spanish and Portuguese possessions obtained in 'the first non-competitive era'. The dominant arena of the 'first competitive era' was the Caribbean (initially, for plundering the Spanish Empire). Subsequently, the Greater Caribbean (from Maryland in the north to north-east Brazil) was to become the principal geographical focus for plantation agriculture supplying sugar and tobacco to the core countries. As the North American colonies did not develop a staple crop, according to Taylor (1985) they effectively prevented themselves becoming peripheralized and also, became the location of the first major peripheral revolt.

The fifteen regions are listed in Tables 1.1 and 1.2 with the period of colonization shown according to broad 50-year bands for the first competitive era and 25-year bands for the second competitive era. As Taylor's map (1985: 67) indicates, these areas covered most of the world, the major exception being China, though even here, the leading core states delimited their 'spheres of influence'.

Table 1.1 Establishment of colonies: arenas of first competitive era

1. Iberian America	1500–1800*
2. Greater Caribbean	1500–1880**/1925
3. Northern America	1600–1800/1850
4. African ports	1500–1850
5. Indian ports	1500–1800
6. East Indies	1500–1925

Source: Taylor, 1985: 82–4.
Note: *Intervals indicated cover creation of colony, reorganisation of territory and, in certain cases, transfer of sovereignty.
**Period of main colonial activity.

Table 1.2 Establishment of colonies: arenas of second competitive era

7. Indian Ocean islands	1600–1900
8. Australasia	1750–1925
9. Interior India	1750–1925
10. Indo-China	1850–1900
11. Interior Africa	1825–1925
12. Mediterranean	1500–1925
13. Pacific Ocean Islands	1750–1925
14. Chinese ports	1500–1925
15. Arabia	1800–1925

Source: Taylor, 1985: 82–4.

Taylor (1985) also provides a brief overview of the economics of formal imperialism, illustrated by two classic cases, the Caribbean, and Africa, and the manner in which they were incorporated into the economies of the core.

The zone from north-east Brazil to south-east North America was converted into 'plantation America', largely to produce tobacco and sugar for the new 'tastes' of consumers in the core region. By 1700, labour-intensive production was met by African slaves with the sugar plantations becoming, according to some interpretations, 'the precursors of the organisation that was to become the factory system in the industrial revolution in the core'.

In Africa, coastal stations were first established for exporting slaves. Drawing particularly on the work of Wallerstein, Taylor (1985) shows that in the final quarter of the nineteenth century, with the colonization of the entire continent, Africa became incorporated into the world-economy as a new periphery, with its space economy divided into three zones: the first, producing for the world market, with each European colony having its own administration and infrastructure to channel commodities into the world-economy; the second, a zone of production for the local market where peasant farmers produced for labour working in the first zone; the third, a large zone of subsistence agriculture, integrated into the world-economy through its export of labour to the first zone.

Taylor's (1985) generalized account, applicable to European imperialisms as a whole, has recently been fleshed out by Christopher's study of the largest of these, that of 'the British Empire at its Zenith' (Christopher, 1988). From this point, the relation between colonialism and urbanism is discussed in relation to the experience of British imperialism, with the implication that the same systemic relations can be established for other core states.

What both Taylor (1985) and Christopher (1988) (who takes a particularly benign view of imperialism) demonstrate is the theoretically well-established, symbiotic and interdependent relationship of the first international division of labour, with peripheral colonies producing primary products and raw materials for the industries of the core and receiving manufactured goods in return (see also Hobsbawm, 1969; Barrat Brown, 1978). While the empirical data is discussed in more detail in the chapters that follow, the broad outline is given as follows.

The shift to urban industrial capitalism at the core is part of the same process as the shift to agricultural and mining capitalism in the periphery. Obviously, at any time during a period of three centuries, colonialism was to account for varying proportions of Britain's trade and other relations with the world-economy in general. Yet as Christopher (1988) indicates (pp. 23, 37, 68), from 1875 the Empire was to become increasingly important so that by 1931, two-thirds of exports, by value, went to British possessions or dominions overseas and something under half of imports came from there; of almost

5

£3,800 million invested overseas in 1914, virtually half was invested in Empire; of the 25 million migrants who left Britain between 1815 and 1924, 10 million went to constitutent parts of the Empire overseas.

It was this spatial division of labour, production, and markets that helps to explain demographic, industrial and urban growth in Britain: between 1815 and 1931, the population grew from 10.5 to 44.5 million, with the urban population quadrupling in the century before 1914; whereas in 1801 there was only one large city with more than 100,000 inhabitants, by 1931 there were 50, in which lived half the population. The factory system that centralized production in Britain depended for many of its basic raw materials (cotton, wool, rubber, tin, and other minerals) on supplies from its colonial possessions, and these production processes were concentrated in specific specialized regions. Food products also came from the Empire, particularly wheat, rice, tea, sugar, etc. Likewise, the colonies depended largely on British capital, shipping, insurance, managerial expertise, as well as cultural products in the broadest sense: education, science, language, religion and also architecture, planning, and design.

Yet where the main economic function of the colonies was the production of mineral and agricultural products and raw materials, and hence, the focus was rural, the manifestations of colonialism were equally urban; first, in the political, administrative, and economic role of its cities and towns in their function of control and surplus extraction; subsequently, in their increasingly signficiant role as markets, centres for consumption and 'theatres of accumulation' (Armstrong and McGee, 1985).

The impact of colonialism on the spatial organization and urban system of colonized countries is well known and need not be rehearsed in detail here: the establishment of externally oriented port cities (whether in Latin America, North America, South and South-East Asia, or Africa) through which goods were exported back to the core and, in the later phases of colonialism, manufactured goods imported to the periphery; the reorganization (or the organization) of urban hierarchies in the colonized societies and the establishment of political, administrative, and military centres, including colonial capitals. It was from among this colonial urban system that a majority of today's 'world cities' were to develop (King, 1990).

Earlier studies of these colonial cities, undertaken within dominant 'diffusionist', 'modernization' or 'stages of economic growth' paradigms of world development of the 1950s and 1960s, were to conceptualize them as examples of 'diffusion', 'westernization' or 'Europeanization'. More recently, theoretically informed empirical research (discussed in Chapter 2) has demonstrated the essentially integrated and interdependent nature of urbanization and urban processes (and in some cases, urban forms) between respective peripheral colonies and metropolitan cores as all were incorporated into a single division of labour of the capitalist world-economy (Wallerstein, 1974: 1984). What these studies demonstrate is, not only can colonial urban development not be understood separately from developments in the metropole but similarly, urbanism and urbanization in the metropole cannot be understood separately from developments in the colonial periphery. They are all part of the same process. It is these issues that are addressed in the following chapters.

CHAPTERS: CONTENT AND THEMES

Earlier versions of Chapters 2–5 were first published in various and disparate places, and for different readerships, in the early 1980s (see p. 173). However, in putting them together here, I have made substantial revisions, not only in bringing them up to date but also, in placing them in a more theoretically coherent context.

As suggested on p. 6, colonial cities were the major links between core and peripheral economies during the period of imperialism, articulating the flow of capital, people, commodities, and culture that flowed between them. Long before the phenomenon occurred in the metropolitan capitals, colonial cities were the sites for the encounter, on any significant scale, between representatives of capitalist and pre-capitalist social formations, between what today we term 'developed' and 'developing' societies and peoples, between racial, cultural, and ethnic groups from Europe and other groups from the continents and regions where the colonization took place. In a very real sense, they were 'global pivots of change' (King, 1985: 7–32), instrumental in creating the space in which today's capitalist world-economy operates.

Chapter 2 first looks briefly at the concept of the colonial city and the way in which it has been understood since it was first seriously considered in the 1950s; then, drawing on theoretical and empirical studies, it suggests its characteristics, as also a typology of European colonial cities and a spatial framework within which their various dimensions can be investigated. Finally, on the basis of various case

7

studies, it discusses the relationship between their function, organization (in terms of governance and social control), and space (and the way this encoded and influenced stratification and categorization); and between social structure, culture, and the built environment.

The interest in post-modernism since the early-1980s (Jameson, 1985) and the work of Edward Said (1978) provide an appropriate backdrop for a revisionist look at the topic of the 'export' of urban-planning ideologies and norms from metropole to colony, first explored in 1977 and later revised (King, 1977; 1980). Jameson suggests that:

> the economic periodization of capital into three rather than two stages (that of 'late' or multinational capitalism now being added to the more traditional moments of 'classical' capitalism and of the 'monopoly' stage or 'state imperialism') suggests the possibility of a new periodization on the level of culture as well.
>
> (Jameson, 1985)

Jameson (1985) goes on to state that, from this perspective, the 'high' modernism of the 'International Style' (and I would add here, particularly in architecture and planning) would 'correspond' to the second stage of monopoly and imperialist capitalism that came to an end with the Second World War (Jameson's periodization here is, however, faulty: the 'Second World War' is a Euro- and Americo-centric notion of historical change that, for previously colonized states, begins with their date of independence, the first being that of India in 1947). The third stage of 'consumer capitalism' 'corresponds' to the 'post-modernist' stage (quoted in Knox, 1988: 2).

Whilst the imperial foundations for, and context of, 'modern' urban planning as well as the 'International Style' were first set out in the original versions of Chapters 3 and 4 (King, 1977; 1983), Jameson's threefold periodization makes the relationships between the stages of capitalist development and cultural forms (especially as they relate to architecture and planning) more explicit. Both in the metropolitan core as well as the colonial periphery, it was the political and economic conditions of imperial dominance that, to a very considerable extent, were the prerequisites of some of the early 'rational' planning experiments in Britain, whether in cocoa-based Bournville (near Birmingham), British India's capital of New Delhi, or the great 'Master Plans' [sic] of Pretoria or Canberra. These,

8

as other examples of 'modern' urban planning, were presupposed on Britain's particular and privileged industrial and urban place in the world-economy, a fact that is generally overlooked (e.g. Relph, 1987). Likewise, Lyautey's designs for Morocco or Corbusier's seven plans for Algiers were founded on the colonial connection (Abu-Lughod, 1980; Wright, 1987; 1991; Rabinow, 1989a; 1989b).

In the formal institutionalization of 'town planning', the notion of 'modernity' and 'the modern' was informed by two sets of circumstances: the first, constructed diachronically, was in relation to the premodern, preindustrial, or early industrial capitalist cities of Britain, to replace the 'disorder' and 'squalor' of the old industrial towns; the second, constructed synchronically, in relation to the 'traditional', 'unmodern' societies confronted in the colonial encounter, especially from the 1870s, and conceptualized in the new colonial science of anthropology (Asad, 1973; Wright, 1987). The new cities, and the new 'norms and forms' (Rabinow, 1989b) introduced from the metropole to the colony did not simply provide 'models on which the colonies built' (Christopher, 1988: 34, an innocent and Eurocentric assumption), they were also the 'norms and forms' of one mode of production (industrial capitalism) being transplanted into the territory of another mode of production, though these were also couched in different cultural forms (i.e. French, British, and German) and expressing particular discourses of power/knowledge, class and professional ideologies of the 'ideal city'. And they were used, both consciously and unconsciously, as social technologies, as strategies of power to incorporate, categorize, discipline, control, and reform, in terms of symbolic codes and new systems of classification, both the colonial as well as the indigenous populations (McGee, 1967; King, 1976; 1984). The social, the racial, and the spatial were embodied in explicit linguistic and conceptual form: 'the native city', 'European hospital', 'Black townships'. The new colonial cities were to become the training grounds of new consumers but also, the sources of resistant cultures and the expression of national identities and new subjectivities.

Whilst some of these issues are raised in Chapter 3 on 'Planning in the colonies', this is still, as it was a decade ago, a much under-researched area with many studies of a normative nature though few focused on the way in which built space and form modify consciousness, change habits of consumption, contribute to new forms of stratification; or the way in which new codes, regulations, and professional practices give rise to different social meanings that are then themselves consumed to change subjective consciousness.

These first two chapters are concerned with understanding changes on the periphery. Yet as indicated on p. 45, these cannot be understood without references to economic, social, and urban developments at the core. Chapters 4 and 5 therefore consider some preliminary issues raised by looking at the historical development of cities within a world-, and particularly, a colonial world-systems perspective. I have made few changes since they were first published in essay form in order to convey, first, the problems as they were originally addressed and second, the argument as it subsequently developed in Chapters 6 and 7, both of which provide empirical evidence on the connections between colonialism, urbanism, and the world-economy (see also King 1990).

As will be indicated in Chapter 4, one of the most influential paradigms within which to consider urban phenomena globally, has been the world-systems perspective (Wallerstein, 1987). The selective use of this concept, combined with theoretical developments in urban political economy in the 1970s, produced the world political-economy approach to the study of urbanism in the 1980s. Yet these approaches have largely been characterized by a materialist orientation to the urban which, till recently, has not addressed the problem of the ideational, symbolic, and cultural, or the realm of the 'pre-economic'.

Robertson (1988) has recently drawn attention to this neglect, emphasizing 'the sociological significance of culture' and providing insights into why it has been neglected. In particular, he discusses the different orientations and conceptualizations towards 'culture' in sociology and anthropology, two disciplines, the separation of which 'is almost entirely attributable to a particular conjuncture in the long term process of globalisation' (p. 14). To oversimplify Robertson's (already simplified) argument, because of anthropology's greater exposure to global heterogeneity, it has been given greater salience to the concept of culture whereas sociology, because of its greater exposure to apparent homogeneity in studying 'modern' societies, has incorporated the use of 'culture' only selectively and on its own terms (p. 10).

It is this tension that is the subject of Chapter 5, interrogating as it does, an earlier 'social anthropological' and culturological' explanation of colonial urbanism (King, 1976) with some of the materialist questions prompted by urban political economy.

The greatly increased sense of globalization in the 1980s has heightened these debates, giving an increased urgency to developing appropriate paradigms in which to understand the development of 'the world as a single place'. Here again, it is useful to pay attention to Robertson's recent work (1988) on globalization theory.

After initially exploring the way culture could be integrated into world-systems theory (Robertson and Lechner, 1985), Robertson has distanced himself from what he sees as overly economistic paradigms, to develop a theory of globalization (Robertson and Lechner 1985; Robertson, 1987). What he brings out is that it is crucial to recognize that the contemporary concern with civilizational and societal (as well as ethnic) uniqueness, as expressed via such motifs as identity, tradition and indigenization, *largely rests on globally produced ideas* (emphasis added). In an increasingly globalized world, 'characterised by historically exceptional degrees of civilizational, societal and other modes of interdependence and the widespread consciousness thereof, there is an *exacerbation* of civilizational, societal and ethnic self-consciousness'. Robertson seeks an empirical basis for a civilizational analysis that transcends and subsumes, older, Western-centred discourses to give greater attention to civilizational and societal distinctiveness. Globalization involves 'the development of something like a global culture, not as normatively binding but in the sense of a general mode of discourse about the world as a whole and its variety' (ibid). More recent discussions by these and other scholars on the themes of universalism and particularism, commonality and difference, the local and the global, equally undermine any simplistic assumptions about a general 'homogenization of culture' (see King, 1991, forthcoming).

In the last few years, there has been a rapid growth of theoretical and empirical studies on the global context of metropolitan growth and on cities and the international division of labour. However, most studies have tended to focus on one 'end' of that spatial division, examining the situation of one particular city. In the issues considered in Chapters 4 and 5, and in the more empirical examples discussed in Chapters 6 and 7, by moving between core and periphery, my aim has been to explore and demonstrate the symbiotic, complementary, and essentially interdependent nature of the development of urbanism in the capitalist world-economy. However, for the development of such globally oriented and historical studies of urbanism, these recent theoretical studies referred to here help to bridge the gap between the world political-economy approaches that have developed in the 1980s and more culturally oriented perspectives that can be identified in the following chapters.

Chapter Two

INCORPORATING THE PERIPHERY (1)

Colonial cities

SOME CONCEPTUAL CLARIFICATIONS

How useful is the concept of the 'colonial city'? The difficulties posed by the use of this idea arise from the radical transformation in the understanding of both the city and colonialism since the concept of the 'colonial city' (that is, of industrial capitalist colonialism) was first seriously considered in the 1950s (Redfield and Singer, 1954). On the one hand, there is the shift in urban theory from a localized interest in urban ecology, through a materialist theory relating urbanism to modes of production in the 1970s, to a pluralist urban political economy of the 1980s that links cities to a capitalist mode of production as well as sociocultural processes operating on a global scale.

On the other hand, the understanding of colonialism and its relationship to 'development', the world-economy, globalization, or cultural change has likewise gone through a series of paradigm shifts in the same period, understood through theories of diffusion, modernization, and dependency in the 1950s and 1960s, and world systems, modes of production, and the internationalization of capital in the 1970s and 1980s (Chilcote, 1984; Preston, 1986). It might be useful, therefore, to look briefly into the history of the concept.

One of, if not the earliest consideration is of the anthropologists Redfield and Singer in 1954. For them, colonial cities were 'the mixed cities on the periphery of an empire which carried the core culture to other peoples'. They were examples of the 'heterogenetic city', the city of the technical order, where 'local cultures are disintegrated and new integrations of mind and society are developed' in contradistinction to the 'orthogenetic city', the city of the moral order, the city of the culture carried forward.

The heterogenetic transitions had grown with 'the development of the modern industrial worldwide economy . . . and . . . the expansion of the West' (Redfield and Singer, 1954). Such cities included the colonial cities of the European powers (such as Jakarta, Manila, Saigon, Bangkok, Singapore, and Calcutta) 'which admit native employees daily at the doors of their skyscraper banks'. Thus, for Redfield and Singer, these cities are products both of 'the modern worldwide economy' and 'the expansion of the West'.

Later formulations refer to the colonial city as representing the introduction of 'Western' urban forms into 'non-Western' countries (Abu-Lughod, 1965) or to the 'cultural hybrid' of the colonial city as subsuming elements of both the 'traditional' and 'modern' world (Wheatley, 1969).

However, in *The South East Asian City* McGee's (1967) substantial section on the colonial city added a new dimension. Whilst the larger context is still provided by 'the impact of the West', his specific reference to a given mode of production (though the phrase is not used as such) distinguishes his account from earlier writings, i.e.

> The most prominent function of these cities was economic; the colonial city was the 'nerve centre' of colonial exploitation. Concentrated there were the institutions through which capitalism extended its control over the colonial economy — the banks, agency houses, trading companies, shipping companies.
> These banks . . . were, of course, largely European owned.
> (McGee, 1967: 56–7)

Or again: 'A situation was created in which the countryside, with the exception of the enclaves of foreign capitalism — mines and plantations — became increasingly impoverished in comparison to the towns' (p. 72).

Whilst McGee elsewhere refers to the introduction of 'Western-type suburbs' and 'Western' city models into South-East Asia (p. 139), not yet specifying that these were themselves products of capitalist forms of development in the core, the attention given to capitalism in the colonial transformation of South Asian cities predates some of the insights from the 'new urban sociology' of the 1970s.

This change in understanding was to become more developed and widespread from the mid-1970s. On one side was the work of Wallerstein and others on the world-system, which restructured perceptions about the development of the world-economy. What Wallerstein indicated was that the present global economy emerged not simply from 'the expansion of Europe' but from 'the expansion of the capitalist mode of production' (Wallerstein, 1976: 30). About the same time, the theoretical work of Castells (1972; 1977) and Harvey (1973) was to develop the connections between city-forming processes and the larger historical movement of industrial capitalism. As Harvey pointed out, 'Urbanisation is economic growth and capital accumulation, and these processes are global in their compass' (Harvey 1975: 99).

Yet while colonialism and the colonial city are seen as stages in a worldwide expansion of capitalism, this process is neither unproblematic, unidirectional, let alone predetermined. It is an ideological and cultural process in itself (King, 1976) as well as one that meets pre-economic cultural resistances (Robertson, 1988). To recognize that the colonial city was an instrument in the expansion of the capitalist world-economy and that it has also led to alternative futures (Forbes and Thrift, 1987) still leaves many questions unanswered. Some of these are addressed next.

In at least two senses, all cities can be described as colonial: at the local level, the powers that form them organize their hinterland and live off the surplus the non-urban realm provides. At the global level, existing cities organize the surplus both of their own society as well as that of others overseas (see also Johnston, 1980: 67–76); the local relationship of town-to-country becomes the metropolis–colony connection on a world scale (Williams, 1973).

Historically, it can be supposed that cities were first created by the exercise of dominance by some groups over others to extract an agricultural surplus and provide services in one geographically defined society. Subsequently, other settlements may have been planted within that society as a means of furthering political control and both creating or expropriating a surplus. The logic of this process is to further extend, by developments in transportation (initially shipping), the boundaries of one society to incorporate other territory and peoples overseas. In this, the city — as a cultural artefact — becomes an instrument of colonization, defined, for our purpose here, as 'the establishment and maintenance, for an extended time, of rule over an alien people that is separate and subordinate to the ruling power' (Emerson, 1968). Our intellectual reservations, therefore, concerning the usefulness of the concept, seem overruled in the face of the existence of a 'real object', though one still dependent for its origins on a process begun elsewhere. In the four and a half centuries of European colonialism (*c.* 1500–*c.* 1950) two of the basic characteristics of these cities were that they were established (with the aid of extensive transportation technology) at a distance from the metropolis and in areas with a totally different economy, society, polity, and culture.

In short, therefore, it may be preferable to speak of a 'city in a colonized society or territory' rather than a colonial city *per se*. This is more than a semantic adjustment: it turns our attention away from the idea of a 'special category of city' to considering at least four elements: a society, the territory and location where it is, the process of colonization, and the city that results.

Yet even in its conventional sense, 'colonial city' is a very broad category, blurring as many features of the reality as it illuminates. At the simplest level, a typology of *all* cities must account for a variety of historical situations where settlements are transplanted by a colonial power, for example, those of the Romans in France, the Arabs in North Africa, the English in Ireland, the Moghuls in India, the Russians in Central Asia, as well as the Portuguese in South America or the Dutch in Indonesia considered in *Colonial Cities* (Ross and Telkamp, 1985). These colonial situations can be classified according to various criteria: mode of production (mercantile capitalism, industrial capitalism); energy base (pre-industrial, agricultural, industrial, with animate or inanimate energy sources); historical time (seventeenth, twentieth centuries); culture and geographic area (Arabic, African, and Asian); society (Dutch, French, as well as strata within these societies); and no doubt others. In the collection of essays for which this chapter was originally written (Ross and Telkamp, 1985), the definition of colonial cities was made on particular 'developmentalist' criteria that included historical colonial cities in Latin America, Africa, South and South-East Asia, but not those in North America or Australasia. With two exceptions, these are cities:

1. Where the dominant colonial minority is culturally European (though ethnically, representing different national groups), racially Caucasoid and (in contrast to the religious or belief systems of the areas where they were established) nominally or actually Christian;
2. of what are now independent states where the earlier colonized peoples have subsequently re-established (or, in the case of the West Indies, established) at least formal sovereignty over their own

territory (the exception being Cape Town). They are, in short, cases of 'limited' or 'unsuccessful' colonization. With the exception of Rio de Janeiro, cases not considered in that account are the cities in other 'successfully' colonized lands where the indigenous population was largely eliminated, marginalized, or to varying degrees absorbed into the population of the colonizing power, and where the colony subsequently became independent (e.g. in relation to Britain, the colonial cities of Boston or New York), or, despite political independence, where strong economic, political, or cultural interests linked the city to the metropolis (e.g. Sydney in Australia or Halifax in Canada);
3. they are cities created or influenced by the emergence of a world system of capitalism.

The inclusion of these latter areas, particularly when considering pre-nineteenth-century cities of the mercantile era, would be both conceptually logical and, for a genuinely comparative understanding of the function and form of colonial cities as a whole, would be methodologically desirable.[1]

SOME SUGGESTED CHARACTERISTICS

Telkamp (1978), in a comprehensive survey of the literature and drawing particularly on studies by Bellam (1970), Brush (1974), Horvath (1969; 1972), King (1976), Lewandowski (1977), Lockard (1976), and McGee (1967), indicated some thirty features that were seen to characterize the colonial city. These can be classified into seven categories referring broadly to the geopolitical, functional, political/economic, political, social/cultural, racial/ethnic, and physical/spatial features of the city. Some attributes have, for obvious reasons, been placed in more than one category.

Geopolitical

1. External origins and orientation.

Functional

2. Centre of colonial administration (some exception to this);
3. Multiplicity of functions, with presence of banks, agency houses, insurance companies, etc.;

17

4. Focus of communications network;

5. Acts as economic intermediary — symbolized by corrugated iron *godowns* (or warehouses).

Political/economic

6. Dualistic economy, dominated by non-indigenes;

7. Presence of large group of indigenous unskilled and semi-skilled migrant workers (that see the colonial city as an alien community);

8. Municipal spending distorted in favour of colonial elite;

9. Dominance of tertiary sector;

10. Parasitic relations with indigenous rural sector.

Political

11. Eventual formation of indigenous bureaucratic-nationalist elite;

12. Indirect rule through leaders of various communities.

Social/cultural

13. Social polarity between superordinate expatriates and subordinate indigenes;

14. Caste-like nature of urban society;

15. Heterogenous dual, or plural society with three major components:
  (a.) Elite formed by residents of colonial/imperial power with externally derived authority based on military force;
  (b.) Intervening groups that originate from racial mixing and in-migration from other colonial or semi-colonial territories (e.g. overseas Chinese);
  (c.) In-migrated indigenous resident groups consisting of educated modern intelligentsia and modernizing elites, as well as uneducated ethnic groups, tribes, clans, etc.;

16. Occupational stratification by ethnic groups;

17. Pluralistic institutional structure;

18. Residential segregation by race;

19. Large groups of unskilled indigenous and semi-skilled migrant labour.

Racial/ethnic

20. Racial mixing;

21. Occupational stratification by ethnic groups;
22. Racial residential segregation.

Physical/spatial

23. Coastal or riverine site;
24. Establishment at site of existing settlement;
25. Gridiron pattern of town planning combined with racial segregation;
26. Urban form dictated by 'Western' models of urban design;
27. Specific character of residential areas;
28. Residential segregation between exogenous elite and indigenous inhabitants;
29. Large difference in population densities between areas of colonial elite and indigenous population, impacting life style and quality of life;
30. Tripartite division between indigenous city, civil and military zone.

Such a list[2] is neither exhaustive nor comprehensive and also reflects the particular urban theoretical models available when the research was undertaken. Nor are these characteristics, when taken individually, unique to the colonial city. The features that distinguish the city are those that Balandier (1951) saw as intrinsic to the colonial process itself, namely: (1) dominance by a foreign minority, racially (or ethnically) different, of an indigenous population, and inferior from a material standpoint; (2) the linking of radically different civilizations in some form of relationship (this is perhaps the special task of the colonial city); (3) the imposition of an industrialized society onto a non-industrialized one (this, of course, applies only from the late eighteenth or early nineteenth centuries); in (4) an 'antagonistic relationship' where the colonial people were 'subjected as instruments of colonial power'.

According to this, the unique features of the colonial city are apparently these:

1. Power (economic, social, political) is principally in the hands of a non-indigenous minority; the rights of the colonized are either nil or very restricted.
2. This minority is superior in terms of military, technological, and economic resources and, as a result, in social organization.
3. The colonized majority are racially (or ethnically), culturally,

19

and religiously different from the colonists who (in the cases considered in the Ross and Telkamp collection (1985)) are culturally European, and by religion, nominally Christian. While other suggested characteristics can be found in 'non-colonial cities' (e.g. cultural pluralism, occupational stratification by ethnic groups, large groups of indigenous unskilled labour), what distinguishes the colonial city is the degree of scale to what the characteristics listed are manifest, and their combination in a particular urban ensemble. They also become of interest when compared to the characteristic attributes of the developing 'world cities'.

(King 1990)

TOWARDS A TYPOLOGY OF EUROPEAN COLONIAL CITIES

It has been suggested that there were basically two types of trans-oceanic colonization: for Christopher (1988: 3) the basic distinction is between (1) European settlement; and (2) the exploitation of indigenous societies. 'Each resulted in the formation of two very different societies and economic systems within the colonies and hence, distinctive landscapes'; for Johnston (1980: 70–2), the same situations are classified by different criteria. In the first type, colonists encounter 'a virtually unoccupied new land' in which they have to establish a population to extract the needed surplus. In the second type, they encounter a 'relatively dense distribution of population in particular societal forms'. Both these formulations leave various unanswered questions, not least for understanding colonial cities, their impact on existing modes of production, or the particular structure they adopted. If the literature on colonial cities has suggested that there was some recognizable 'ideal type' in the late nineteenth and early twentieth centuries, there is far less agreement about the existence of a common type before this time.

To understand colonial cities we need a typology that accomodates a large number of criteria. We must first distinguish between societies and territories on the basis of the number of inhabitants, and the nature and level of their economy and culture, including the existence — if any — of urban settlement. We might for example, distinguish between colonized societies or territories which:

1. have a highly developed culture and a long (and flourishing)

urban tradition, for instance, India, with Benares, and Nigeria with Benin;

2. have substantial numbers of indigenous inhabitants with a self-sufficient (usually peasant) economy but few, if any urban settlements, for instance, South Africa and North America.

3. have either no, or very few indigenous inhabitants whose level of economic and social development is relatively simple and who have no permanent forms of settlement, for instance, Australia;

4. have had a sophisticated and developed urban tradition that has subsequently declined prior to colonization, for instance, the Aztec civilization.[3]

A second set of criteria needs to take account of the motives for and circumstances of colonization. These would include a variety of circumstances from organizing an existing surplus through simple trading, enforced trading, creating and exporting a surplus through colonization, or strategic colonization for political rather than economic purposes, and religious conversion.

In the third place, the motives and circumstances obviously govern the numbers and degree of permanence of the colonizing population. These, in turn, affect the degree of coercion exercized over the indigenous inhabitants. This might range from none (where there are no inhabitants), to some (where conquest is made by treaty or the forced leasing of land or labour), to the complete enslavement or, at the extreme, annihilation of the population.

Linked to the question of motivation is the resource being sought and how this is determined by changes over time in the technological and cultural appraisals of the metropolitan power. 'Resources are technological and cultural appraisals . . . their quantity is dependent upon individual preferences existing in the population and the cognitive skills which people possess to help them exploit the resource system' (Harvey, 1973: 69). These might be silk, precious metals, land, people, copper, tea, coffee, or, with the invention of the internal combustion engine, oil, rubber, and most recently, uranium.

As to the physical, spatial (and social) form that the city developed, at least ten possibilities affected the outcome. Thus, where an indigenous settlement already existed, the colonists have the following choices, depending on their intentions, numbers, requirements, and the stage of colonization:

21

1. The site and accommodation is occupied with little or no modifications (Zanzibar).

2. The site and accommodation is occupied but modified and enlarged (as with many small inland colonial administrative centres).

3. The existing settlement is razed and built over (Mexico City).

4. The site and accommodation are incorporated into a new planned settlement (Batavia).

5. The new settlement is built separate from but close to the existing one (New Delhi).

6. The existing settlement is ignored and a new one built at a distance from it (Rabat).

Where there is no previous indigenous settlement, the new foundation can be built:

1. for the colonists only; non-colonists (i.e. indigenous inhabitants or 'intervening groups')[4] remain outside, providing their own settlement and accommodation (New York);

2. for the colonists only; no other permanent settlement by non-colonial groups is permitted (Sydney);

3. for the colonists but with separate location and accommodation for indigenous and intervening groups (Nairobi);

4. for the colonists and all (or some) of the intervening and indigenous groups in the same area (Kingston).

Each of these situations can be modified according to whether the colonial settlement was planted on the basis of a preconceived plan or whether it simply grew by accretion.

FRAMEWORKS FOR ANALYSIS

What insights can be derived from studying colonial cities? The answer depends on one's perspective: studying the colonial city is not about developing a model but about understanding processes on a global scale. From a 'development' viewpoint, it provides the context for testing theories of dependence, or the emergence of global capitalism. It is a laboratory for testing hypotheses: for geographers, on the cultural variable in spatial change; for anthropologists, on the 'Westernization' of material culture; for sociologists, it poses questions about the degree of universality (and transferability) in institutions or social processes that the 'artificial' establishment of colonial societies provides an ideal laboratory to investigate; for architects or planners, it demonstrates the distinction between notions of a 'cultural order' and the 'rational professionalism' of the Western capitalist city.

My own interest is concerned with these and other issues: what can we understand about a society by examining its physical and spatial environment? And conversely, what can be understand about the physical and spatial environment (the buildings, the architecture, and the spatial structure of the city) by examining the society in which they exist (King, 1980). To answer these questions, however, we first need to know the functions of the city, the form and working of its institutions and organizations, the distribution of power between different groups, and something of their values, behaviour, and activities. And where this chapter perhaps emphasizes the physical and spatial dimensions of the city, it is clear that an understanding of this, as all other aspects, requires a thorough knowledge of the economic and political history of each city, including the social relations of production.

Whatever our perspective, some kind of framework is necessary within which to examine and compare a variety of colonial cities at different historical times. Urbanism, to quote Harvey again, 'is not just the history of a particular city but the history of a system of cities' (Harvey, 1973: 250). It is, indeed, even more than this, for the system exists at various levels. Which of these levels we choose to examine, and the order we accord to them, will depend on the problem to be investigated.

1. The city in relation to the colonized society or territory (see Figure 2.1).
Here we can consider the existing indigenous economic, social, political, and cultural forms or local conditions (resources, climate, and environment) that contribute to the distinctive character of the colonial city: the Indian contribution to Calcutta, the African contribution to Lagos; the effect of the environment on settlement patterns or emerging culture At this level, the city can be viewed as a spearhead of economic, political, and cultural penetration, following which the structural organization, or reorganization, of the colonized society or territory takes place. Such organization includes the reorientation of trade and transport; reordering the urban hierarchy and establishing a new system of towns; the

23

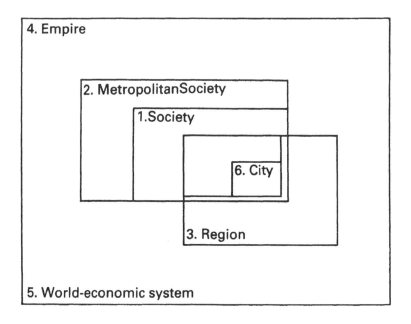

Figure 2.1 Framework for investigating aspects of the colonial city

emergence of new occupations and systems of stratification; the creation of new bases of political power, the growth of new elites; the promotion of cultural change, in religion, education, science, language — including 'Westernization' — and resistance in the politics of cultural nationalism; the direction or redirection of the economy to a metropolitan and world system and through this, the development of a labour market in the colonized society that affect both the colonial city as well as rural areas, with consequent agricultural decline; demographic change, particularly migration, with tensions between city and rural areas, between (where they exist) 'traditional' and 'modern' cities, and involving the break-up of kinship and tribal structures with consequent social disorganization, or the distinctive social, ethnic, or racial composition of the city.

2. The city and the metropolitan power.

At this level are considered those factors that influence the colonial society and city itself: switches in capital investment from domestic to overseas; booms and slumps affecting emigration; ideological changes, or shifts in political power in the metropole motivating people to the colonies; changes in metropolitan colonial policies, particularly as they impact urban and regional development (Dossal, 1989); colonial policies affecting indigenous and colonial subjects; metropolitan attitudes in regard to indigenous cultures, including racial, social, or legal issues, etc.

Colonial cities also provide insights into the metropolitan society, its institutions and culture in a way that an examination of metropolitan society alone does not permit. Metropolitan institutions, lifted out of their social, cultural, historical, and, not least, environmental context, and transplanted to colonized — often 'tropical' — lands, can be seen, if not like 'flies in aspic', nevertheless in a new light. Together, they form an interacting urban system (Chapter 7), which also has a hierarchy with the Imperial capital at its head (Christopher, 1988, points out that in 1931, London was five times larger than the next largest city in the imperial hierarchy — Calcutta). Some institutional forms and functions are shared by all the cities in the hierarchy (e.g. certain legislative, administrative, or judicial instruments, professional practices, language, bureaucratic procedures); others are confined to the metropolis (political decision-making at the highest level, principal banking and financial functions, cultural accumulation in museums, or cultural constitution in universities, libraries, or research institutes); other institutions and practices are confined to the colonies (e.g. indentured labour, slavery, particular forms of land tenure, racial segregation by residence, etc.). Functionaries (colonial governors, inspectors of police, engineers, missionaries, and educators) move between the city and the metropolitan power as well as within the colonial urban system itself.

Within the historic British Colonial Empire, for example, there are elements common to London, Madras, Cape Town, Nairobi, Kingston, Halifax, and Gibraltar. This is the appropriate level to examine movements of labour, capital, images, ideas, or goods. It provides the framework within which people and ideas move (e.g. from Britain, via India, to South Africa or Australia) or to understand particular cultural phenomena (the 'Bengal Room' in the 'Victoria Hotel' in 'Vancouver', 'British Columbia', 'Canada').[5]

Specific imperial urban systems (French, Dutch, British, and

25

Spanish) manifest different ethnic and racial characteristics and these, to varying degrees, are changed as cities are fully incorporated into the world-economy. Concepts of government, legislation, religion, language, medicine, or art invite a threefold comparison: with those of the indigenous society, with those of other colonizing powers in the same region, and with those of the metropolis itself. Once exported, metropolitan institutions and behaviour tend to ossify under the influence of 'nativism', a state of mind brought about by the perception of other cultures as inferior; practices long abandoned in the metropolitan society persist and develop (the 'dressing for dinner in the African bush' syndrome). Hence, in understanding colonial societies it is clearly relevant to relate institutions and behaviour to contemporary developments 'back home'. Not least, colonial cities in the colonies have been significant in the development of 'colonial cities' in the 'metropolitan' society of today: 'Ugandan Asians in Britain' is not merely a tripartite geographical expression.

3. The city in the region.

This level might be represented by one geographic continent, either in whole or in part (India, South America, and West Africa) or a larger area (South-East Asia and the Caribbean). Here, different colonial and local powers are operating; in the analysis of the form and function of the city, its demographic composition, the manner of the incorporation of labour, the development of trading relations, or the effects on the colonial city of rivalries between competing colonial powers, this is a significant level of analysis (Basu, 1985).

4. The city as part of empire.

Within the empires — Dutch, Spanish, British, French, etc., understood as economic, political, social, and cultural systems, are different cities. Some of these are 'colonial' in the sense considered here (e.g. Kingston); others are metropolitan (London, Liverpool, and Bristol); others subsequently become independent (e.g. Adelaide).

5. The city and the world system.

Viewed in this context, the colonial city provides an additional key to unravel the complexities of an increasingly evident world-economy. Within this framework, between the 1950s and 1980s, the city (e.g. Hong Kong, Singapore, Jakarta) has been conceptually transformed from the colonial city to the Third-World city to the world city (Redfield and Singer, 1954; Dwyer, 1974;

Drakakis-Smith, 1987; Friedmann, 1986). In this framework, we can consider the role of the colonial city in incorporating the colonial state into the world-economy, or the effects on the city of global developments (e.g. demand for colonial products, capital circulation and investment, long wave periods of expansion and decline in the world economy, wars, etc.).

6. The colonial city *per se*.

These various frameworks, implying different sources of structural influence on the city and its relation to others, do not preclude analysis of the internal dynamics of the city itself, nor of the role of particular agents or municipal politics. Moreover, the frameworks and the phenomena to which they relate are obviously interconnected. Yet because of its distinctive role in incorporating the colonized economy into that of the metropole, in linking one (and often more than one) culture with another, and embodying the characteristic power structure inherent in colonialism, the colonial city has distinctive functions and features, as well as organizations and institutions and these are often represented in the physical and spatial form of the city. The most widespread of these is 'economic dualism' (Santos, 1975), and the 'dual city' in which this was expressed (though these must be understood as interdependent and not separate entities); another is cultural pluralism where race combines with other criteria of stratification (occupation, wealth, and religion) to produce a distinct ecology. The consciousness of race, and racial conflict with which it is often associated, is perhaps the major urban manifestation of colonialism.

FRAMEWORKS FOR INVESTIGATING ASPECTS OF THE COLONIAL CITY

Function, organization, and space

Ross and Telkamp (1985) suggest three concepts for studying the colonial city: function, organization, and space. To these I would add three more, either refinements of, or additions to them: culture, in the sense of a set of rules, models, a way of defining, interpreting, and symbolizing the world; to the concept of space (referring to spatial order, location, distance, and area or 'territory'), can be added built form and built environment. This comprises not just architecture with connotations of design and appearance, but all that is built or constructed; to organization, social organization or more particularly, social structure. Organization may be used to describe the instruments of the state: government, administration, and social control; social structure, to refer to the ethnic, racial, occupational, and socioeconomic composition of the city. Finally, we might add the concept of situation or situational to refer to the particular historical conditions and circumstances prevailing at one period of time, for example, the particular transport technology and communications prevailing in eighteenth-century Batavia; the particular set of beliefs concerning the understanding of disease that, for example, result in the establishment of the hill station Baguio, in late-nineteenth-century Philippines.

On the assumption that the onset of industrial capitalism is the main watershed in the transformation of colonialism, my comments are divided loosely between the pre-early nineteenth century and after that time.

Function

Regarding function, the first point to be made concerned the motivation for and nature of colonization. Here, the principal difference in the sixteenth and seventeenth centuries is apparently between colonial powers such as Spain and Portugal where, complementary to the economic function, are cultural and religious motives of Hispanicization and Christianization. Though South and Central America provide the obvious cases, the issue is more clearly demonstrated in South-East Asia by comparing the activities of the Spanish in the Philippines to the more restricted economic and military activities pursued by the Dutch, the British, and the French (Reed, 1978). In the Ross and Telkamp (1985) volume, Van Oss brings out the autonomous nature of inland colonial urbanization of the Spanish in Central America with the establishment of an autarkic colonial society. Elsewhere (1980), he demonstrates a somewhat different situation in New Spain. In the mid-sixteenth-century Philippines many early efforts to establish a viable settlement in a swidden, subsistence economy met with near disaster before the Spanish conquerors could first create, and then extract, even a minimum surplus; much of the initial profit from Manila was the result of encouraging a profitable galleon trade in the region. However, Reed's detailed study brings out the importance of the hitherto neglected cultural and religious dimension to colonization (see Reed 1978, Conclusion).

This is worth emphasizing for a number of reasons: the religious (specifically Catholic) component in colonization was an ideology that legitimized it; as an institution, the Church also provided, through its hierarchy and religious orders, major instruments of colonization and social control — churches, schools, monasteries, convents, colleges, and hospitals. Van Oss (1980) reminds us of the large number of churches or the extent of land occupied by convents in proportion to that of administrative buildings. Moreover, awareness of this Christian-Catholic element helps make explicit the Protestant nature of other colonial regimes such as the Dutch or British, too often seen as 'secular'. This is not only important in influencing the nature of particular institutions and social arrangements (the poorhouse, or orphanage in Zeelandia: see Oosterhoff, 1985) but, especially in the nineteenth and twentieth centuries, missionary activity and its effect on education, medicine, welfare, and cultural transformation as a whole. We may mention, for example, its relation to the Bengal Renaissance or the growth and development of nationalism in Africa. No less important is the complex Weberian issue concerning the relation of the 'Protestant ethic' to capitalism. Without provoking a sterile debate on the relation of religious/cultural variables to economic development (an unfashionable topic these days) it is none the less evident that any comparative study of ex-colonial cities (e.g. Karachi, Madras, and Lima) and their development, which ignored the religious-cultural variable, would obviously be partial.[6]

This said, however, compared to the many traditional religion-based, pre-colonial cities of South Asia,[7] the rise of the colonial city was not only to shift the locus of power but, as with the move from Poona to Bombay or Madurai to Madras, to shift it from a religious to a secular basis (Kooiman, 1985; Lewandowski, 1977).

Motivations for colonization, therefore, clearly affect the resulting settlement, most fundamentally in regard to whether this was seen to be temporary or permanent. If the latter (as with much of Spanish and Portuguese colonization), the premeditated planned city was the result. Here, the city and its institutions were a major instrument of colonization. Though the origins of the distinctive grid pattern are still disputed, Philip II's 'Prescriptions for the Foundation of Hispanic Colonial Towns' of 1573, incorporating earlier experience, demonstrate this (see Reed, 1978, Appendix). In some cases, the indigenous city was eliminated; in others it was incorporated into the plan; both these instances possibly demonstrate a total and symbolic dominance by the incoming religion and culture (Van Oss, 1980; Benevolo, 1980).

29

On the other hand, where the intention was primarily trade, depending on the degree of coercion and the activities of other powers in the region, a variety of forms, social, economic, and physical might result: a mere landing stage and warehouse, to a 'factory' (itself a fascinating sub-urban form comprising economic, political, social, and military institutions) to a substantial 'port and fort'. A carefully documented comparative account of the actual *buildings*, their uses and inhabitants (for troops, storage, administration, transport, residence, slaves, etc.) and the use of those of the indigenous population, would tell us much about the function and organization of the colonial city (King, 1980; 1984).

The economic functions of the city were obviously related to the type of resource being exploited or traded: precious metals, minerals, silks, spices, timber, slaves, cotton; in turn, this was related to the nature of labour power employed. In this context, the variety of methods makes it well-nigh impossible to generalize about some 'ideal type' of social structure and stratification; labour might be supplied by the coercion of the existing indigenous population (Manila), slavery, whether of an indigenous or imported population (the West Indies and Brazil), indentured labour recruited elsewhere (Mauritius), local free labour employed under normal market conditions (Calcutta), or attracted in from elsewhere (such as the Chinese in Singapore). Later, metropolitan labour might be imported under duress (convict labour in Australia) or under normal market conditions (emigrant English or Irish labour in South Africa); other systems, such as the *encomienda* of Spanish colonialism, were also possible.

A basic question with regard to function is how far the colonizing power actually creates, stimulates, distorts, or destroys an indigenous agrarian or craft economy. Connected to this are questions of cityward migration and the relation of the colonial city to its hinterland. In discussing Calcutta, Marshall (1985) suggests that the city was not only to activate the economy of Bengal but also large parts of western India without greatly undermining the existing economic structure or changing the social relations of production. A century later, however, Bombay was to act as a drain on the rural economy and to create an urban proletariat (Kooiman, 1985). In general, the primate-city phenomenon was a later development, the result of nineteenth-century metropolitan changes.

The extent to which rural peasant economies were destroyed is well illustrated by Abu-Lughod's work on North African colonial cities, though again, this is a nineteenth- and twentieth-century phenomenon. With the establishment or colonial revival of places like Algiers, Tunis, and Casablanca, traditional inland cities on ancient trade routes (Meknes, Fez, and Tlemcen) were undermined. The appropriation and exploitation of agricultural land in Algeria or Libya for the production of Europe-destined products dispossessed peasants who (as in the late-eighteenth-century enclosure movement in England) were sucked into the city to emerge as factory hands (the capital supplied from Europe), dock workers, petty traders, and servants for the European elite. In the twentieth century they were to fill the *bidonvilles*. Ancient Islamic centres such as Quairuwan or Fez, were bypassed by the imposed economy, their economic base in hand production undermined by imported European manufactures, the result of the more total incorporation of the society into the capitalist world-economy (see also Chaichian, 1988). In such classic cases, colonialism extended uneven development, the domination of colonial ports as primate cities, with ecologically different zones for different populations and above all, an excessive expenditure on the 'modern' or 'Western' sector of the city (King, 1976; Rabinow, 1989a), while the poor and rural crowded into the old, decaying indigenous city (Abu-Lughod, 1975; 1976; 1978).

In the Caribbean, the society was not 'influenced' by Europe but actually created by it (Cross, 1979: 9); the plantation economy was an appendage of the metropolis, Kingston, a prime example of 'dependent urbanisation' (Castells, 1977: 47–8; see also Clarke, 1975; 1985).

Organization

We might comment briefly on two aspects of organization: state institutions of government and social control; and some aspects affecting social structure in the city. A basic question concerning the development of institutions of urban governance relates to whether we are dealing with one or two (or more) communities; here, spatial segregation is inextricably tied up with organization. In some cases (e.g. Spanish or Portuguese colonies) there is a higher degree of integration or incorporation in the society. Elsewhere, separate 'European' and 'native' towns have dual sets of legislation, institutions, standards, and provision (by far the most sophisticated system of this is, of course, in contemporary South Africa). In these cases, racial and cultural divisions were institutionalized through legislation and the separate provision of (some) facilities.

31

In the seventeenth and eighteenth centuries, it was apparently sufficient to transplant, with some adaptation, institutions and practices from what were essentially pre-industrial and pre-capitalist European towns to pre-industrial colonial settlements, as accounts of Batavia and Zeelandia demonstrate (see Ross and Telkamp, 1985). What is of interest here is to know the extent of adaptation required by new social, economic, political, or environmental conditions. In British colonies and probably others, the change comes largely in the nineteenth century and stems essentially from the new systems of knowledge and methods of coping with what are defined as 'urban problems' generated by massive capitalist industrial urbanization. This is discussed on p. 38. The extent of state involvement in colonization or whether it was conducted primarily by a trading company (the British or Dutch East India Companies) is also important in affecting administrative institutions, as is the degree to which members of the indigenous society were involved in government. In comparison with non-colonial cities in the metropole, three groups seem to have been especially important in the administration and governing of the colonial city: the military, the church, and mercantile interests.

The development of urban institutions in the cities of different colonial powers serves to highlight the varying cultural provision. Thus, what characterizes Latin American cases (see Ross and Telkamp, 1985) is the way in which religious institutions (particularly Catholic religious orders) provide for education and welfare. Where the Chinese formed a large proportion of the local or 'intervening' population (as in Zeelandia or Shanghai), the family or larger kinship group assumed responsibility for the control of crime, mental illness, the care of the sick or destitute, and also for business finance. This latter was also the case in Rio or Calcutta. In India, certain tasks (like sanitation and the cleaning of public areas) were accomplished through the caste system, an indigenous institution that could be incorporated for municipal purposes.

What a comparative study of the colonial city would probably illuminate is the gradual transfer of functions from the private cultural realm to that of the public sector of municipality or State. In general, the extent of State intervention in the colonial process is extensive, far more than in the metropole (Christopher, 1988). However, the State's relationship with capital, particularly early multinationals (Dunning, 1983) and trading companies (e.g. the Royal Niger Company), needs to be noted.

These relationships, and also institutions of government and social control, are evident in the built environment. The first building to be erected ('the first requirement of a civilised society' according to one colonial administrator) was usually the prison (Christopher, 1988: 89) and those for maintaining order: the fort, the barracks, the court house, the town hall, and the police station. Later, institutions of administration and social control, developed by the centralizing State in the metropolitan society, are transplanted to the colonies, appear as built form in the city — the lunatic asylum, the hospital, and the chamber of commerce, bringing new categories of consciousness and a new social and moral order. Finally, in the reconstitution of colonial cultures, schools, colleges, museums, art galleries, and research institutes appear (King, 1982). With banks and shops, new forms of retailing are introduced and with the critically important institution of the hotel, a space for acts of either social inclusion or exclusion, or for changing habits of consumption and representation.

The club, ostensibly a social institution in British colonial society (King, 1976: 87-8, 172-5) may also be seen as an important mechanism for colonial urban government. In the Chinese treaty-ports such as Shanghai, there developed the *hong*, a distinctive form of organization (of European merchants, taipans, and clerks), which is contained in an equally distinctive spatial form, an archetypical example of the 'colonial third culture' (King, 1976). In short, a whole new set of spaces permitted changed subjectivities.

Some aspects of social structure

The central social fact of the colonial city was its newness, in terms of the colonial incomers, the indigenous people, the migratory groups it sucked in, and the resultant population it generated. All sections were transformed — economically, socially, culturally, politically, and, not least, biologically. The result is much more than a sum of the parts, and over time, the social (and political) structures were massively changed.

Whilst the incorporation of the city into an increasingly pervasive capitalist world-economy meant that, for the later colonial city, economic and social changes begin to make a class analysis more viable, for obvious political and cultural reasons, for much of the time it is race, ethnicity, and gender which are the main criteria for division: the breaks in the society are vertical rather than horizontal.

One insight into the social structure of the early colonial city is through a study of its principal institutions — government, law, economy, religion, social control, kinship, and education (King, 1976, Chapter 3). Clarke (1985) puts emphasis on cultural pluralism, though the degree of cultural assimilation varies greatly both between different colonial societies and over time. In many cases we are dealing with four different 'sets' of cultural institutions: those transplanted, with modifications, from the metropolis; those of the indigenous culture; those of Horvath's (1969) 'intervening groups'; and finally, 'new' institutions that arise from the peculiar demands of the colonial situation (e.g. to mediate between colonizers and colonized, and others of the 'colonial third culture') (King, 1976, Chapter 3); such institutions are present in neither the metropolitan, indigenous, nor 'intervening' society. We might quote as an example here, the institution of 'science', and especially, 'tropical medicine' or 'tropical architecture', comprising ethnoscientific ideas about disease, cultural expectations of health, perceptions of climate and environment, and cultural beliefs and practices regarding various populations in subordinate or superordinate positions. Other distinctive institutions of colonialism are expressed in the built environment as they are in the economic and social relations of production: the plantation, housing for 'single native labour' (Rhodesia), the slave churches of Rio, the *hong* in Shanghai, the bungalow-compound in India, and, of course, the colonial city itself.

In India (and probably elsewhere), the significant factor is not only cultural pluralism but social pluralism: in addition to the 'new' social structure introduced by European colonists based on power, race, social prestige, wealth, and official rank, there was a parallel system recognizing caste criteria, birth, and religion (between Hindus, Muslims, Sikhs, etc.). At different times and places, either or both of these systems might operate. Though Marshall (1985) suggests that by the late eighteenth century there had been little cultural change in the indigenous urban population, a century later, substantial 'Westernization' (in language, education, subjectivity, material culture) had taken place as well as increasing commodification. The manufacture and importation of Western consumer goods helped to give symbolic expression to an emergent class structure (Furedy, 1979; King, 1984: 50–63).

The distinctive social characteristic of the colonial city, however, is the fact of race. As others have pointed out (Swanson, 1969; Rex, 1973) much of the idea of race as we understand it, is essentially an urban product, and very much a colonial urban product. Though race has a biological basis and geographical location, it is also socially constructed. Societies, and races, reproduce themselves. It was in the colonial city where races were in permanent contact, and in contact, were defined in contradistinction to each other (see Mitchell, 1966). Where intermarriage and racial mixing was the rule rather than the exception (as in Latin America) social stratification was of a different nature. Generally, however, race became the characteristic criterion of stratification. A comparative study of the ethnic and racial composition of the major cities of the world from say, the sixteenth century (including the smaller ex-colonial ones), would demonstrate the impact, in visible human terms, of the emergent capitalist world system on the social composition of cities.

A further distinctive demographic characteristic of the colonial city is the relative absence of women and, as far as the colonial community was concerned, of European women. This absence, and from the later nineteenth century, subsequent presence, affected many aspects of colonial urban life most obviously in the creation of mixed-race populations. It also led to specific colonial institutions — the bachelor 'chummery', partly, to the club, and the allocation, in the planned colonial city, of disproportionate space to the provision of recreational activities for the young male official (the race course and riding track). The presence of large numbers of male troops also had numerous implications: in India, high rates of venereal disease led not only to elaborate legislation and the special provision of hospitals and dispensaries but also of vast areas of space laid out for recreational use: football, rackets, polo, gymnasia, and reading rooms in the cantonment, as well as extensive physical space separating this from the local bazaar. The hill station, with its army cantonment isolated from 'native women' on the plains, was a further consequence (King, 1976, Chapter 7; Ballhatchet, 1979).

Space

It is sometimes assumed (e.g. Scargill, 1979) that rigid spatial segregation, on racial criteria, was not only the most notable feature of the colonial city but also, that it was both unique to it and the result of European colonialism. These assumptions need more careful examination. Ancient Indian cities were segregated according to occupation and caste; others in pre-colonial Africa also had ethnic and racial segregation (see Spodek, 1980: 255; Hull, 1976: 80–4; Rabinow, 1989a).

In the colonial cities of Latin America, indigenous Indians lived interspersed with their European colonizers. Both Karasch (1985) and Clarke (1985) note that race, in itself, was an insufficient criterion for spatial segregation. Though spatial segregation according to race, culture, occupation, and socioeconomic status was indeed a characteristic feature of many colonial cities in Asia and Africa in the late-nineteenth and twentieth centuries, a number of comments are apposite.

1. In pre-capitalist European cities, land use was assigned on social and functional, not economic criteria; rigid spatial segregation of social groups was not the rule. Where spatial segregation did exist it was expressed vertically, in buildings, rather than horizontally. This was apparently also the case in traditional Islamic cities.

With the emergence of capitalism a free market in land (now treated as a source of income) developed. With industrialization and the development of a class system from the early nineteenth century, the social structure in European cities became (with the development of new forms of transport) increasingly expressed in the spatial and built environment (Vance, 1971). It was the assumptions and expectations behind these developments that were transferred to the colonial city of the nineteenth and twentieth centuries.

2. The degree of racial, social, and cultural segregation and the extent to which this was spatially expressed in the colonial city was influenced by the practices of the indigenous community (e.g. caste in India) as well as those of the colonizers.

3. The type of spatial segregation was obviously affected by the circumstances and motives of colonization discussed on p. 21. The ideology behind it, whether it was to be permanent, what the relations were to be like with the indigenous or intervening populations affected both the system of social stratification and its spatial expression.

4. The nature of social and spatial segregation changed over time. Between the seventeenth and nineteenth centuries in India, for example, varying numbers of Europeans settled and lived in indigenous areas (for Delhi, see King, 1976, Chapter 8; also 1984; for Madras, see Neild, 1979).

Residential location was governed by the need to live near warehouses, the port, or the indigenous elite. The strict segregation of the late nineteenth century and later resulted from a variety of reasons: personal and military security, especially after the 1857 uprising, genuine as well as perceived health hazards, racial prejudice, cultural preference, social status, and the unequal distribution of land and wealth. From the late nineteenth century, racist and ethnomedical explanations of disease 'rationalized' racial residential segregation (Curtin, 1985; King, 1976; 1984; Frenkel and Western, 1988).

5. The development of a land market and the introduction of the technological infrastructure of the Western industrial city (e.g. underground sanitation, gas and electric lighting, transport systems, and road networks, as well as land use and building regulations) had obvious spatial implications. Here, the distinction between the European settlement and the indigenous city was as much between the modern industrial capitalist and the pre-industrial city, or between the culturally 'Western' and 'non-Western' life styles as it was between the colonial and the colonized (see Abu-Lughod, 1975; Lewcock, 1979). Segregation by race, nationality, or culture was enforced in some places by legislation; it was also, however, a function of political or cultural domination, and especially, the uneven distribution of income.

6. It also seems true that only in a relatively open, mobile society where status is not ascribed, that spatial segregation becomes an index and expression of social stratification. In closed, socially immobile societies, where everyone 'knows their place', physical propinquity is no threat to status. Clarke (1985), for example, suggests that only after the removal of the institutionalized inequality of slavery did marked spatial segregation develop in Kingston.

Later colonial modifications

If there are reservations about treating the colonial city as a single, unitary category prior to the nineteenth century, they multiply after this time. Though we may keep one eye on an individual city, the other must be on colonial urbanization as a whole and the rapidly growing capitalist world-economy of which it is a part.

Emergent industrial capitalism transformed colonialism, whether in the formalization of empire in Asia, the partition of Africa, or the incorporation of Latin America (see Roberts, 1978a). From the later part of the nineteenth century, Wallerstein's concept of the world system as 'a unit with a single division of labour and multiple cultural systems' has increasing heuristic value (Wallerstein, 1979: 390).

At the urban level, where the 'typical' (if not the only) colonial city was previously a port (Basu, 1985) with the development of railways, it became increasingly part of a larger system, of administrative and military towns (district, regional, and national), transport nodes, harbours, mining and market towns, plantation villages, hill stations, and resorts. As these grew, they had their own impact on the existing space economy with regard to domestic industry, agriculture, and the distribution of the population (see especially Abu-Lughod, 1976; 1978; also Saueressig-Schreuder, 1986; Chaichian, 1988). An understanding of this urban system, whether the whole or the smallest part, requires an understanding of metropolitan urbanization and the world and imperial political economy behind it (King, 1976: 27).

Industrialization is a process that uses new forms of energy, technology, as well as capital, labour, and economic and social organization. Urbanization in the metropolitan society grew from this process and within a particular international division of labour such that it relied on the colonial periphery for many of its raw materials. What I would posit, therefore, is the gradual development, within an increasingly global economy, of a transnational capitalist-industrial urban culture. This overlays, or replaces, the more culture-specific pre-industrial, and pre-capitalist cultures of the mercantile era. First manifest in the urban forms and concentrations in Britain in the early mid-nineteenth century, it then develops in Europe and North America and, in the late nineteenth and twentieth centuries, in the colonial and semi-colonial areas of Asia, Latin America, and Africa. Today, it transforms both the traditional and new cities of the Middle East (Johnston, 1980; King, 1984). The main exceptions to these global patterns is the urban culture of state capitalism or socialism (see Giese, 1979). What are the elements of this global urban culture?

The 'modern' capitalist city was both a product of and a help to produce a science-based technology enabling unprecedented numbers to be accommodated in dense settlements: medical science controlled disease, engineering supplied pure water and sewage systems, new forms of transport and communications enabled people to be moved, fed, and employed on a massive scale. New materials and construction techniques (iron, steel, concrete, multistorey buildings, and prefabrication) all helped to produce a form of city based on particular forms of energy and assumptions about the accumulation of capital (Harvey, 1985).

Linked with these new technologies were the institutions, social technologies, and codes that controlled them: with water and health, the engineering and medical professions; with transportation, new legislation and municipal bodies to carry it out; there developed a whole range of social mechanisms and professional knowledges.

From the second half of the nineteenth century, this system of science, technology, urban administration, and ideology became increasingly installed in 'modern' industrializing states in the core, spread via the press, the development of international transport and communication, professional institutions, international conferences, and the like.

European imperialism and colonialism was the vehicle by which this 'modern' capitalist urban culture was transferred to the 'tropical' world as the colonial and metropolitan cities were increasingly linked in a single urban system. In the emerging international division of labour, urban development in India was geared more closely to the colonial needs and policies of Britain, particularly with regard to the supply of cotton and to provide markets for its manufactured goods. From the 1860s, Bombay was replanned and redeveloped, using the urban technologies and knowledges from industrial Britain (Dossal, 1989); later, metropolitan concepts were utilized in the construction of 'working-class housing' (Conlon, 1981; see also Simmons and Kirk, 1981); between 1852 and 1886, the space economy of Burma was restructured, a new urban hierarchy established with Rangoon chosen as the colonial capital. By 1870, the Delta region of lower Burma had become the major supplier of rice for milling centres in London and Liverpool as well as Hamburg and Bremen (Saueressig-Schreuder, 1986); between 1885 and 1900 British imperial power in Egypt had doubled the volume of cotton exports from the country and set up Cairo as the 'colonial city' of control. Between 1897 and 1907, the number of foreign immigrants (mostly European, and colonial functionaries or speculators) in Cairo doubled to reach 16 per cent of the population; in Alexandria it was 25 per cent, and in Port Said, 28 per cent. 'This considerable influx of foreigners divided the city into two parts in a peculiar way: the decline of pre-capitalist Cairo (along with the dissolution of the guilds) was compensated for by the relative growth of colonial Cairo' (Chaichian, 1988: 32).

In India, after the uprising of 1857, the British presence in Delhi and Lucknow remodelled both the social and spatial structure of the city. Railways were introduced, together with a town hall, a clock tower, a new market, and a shopping arcade where European retailing practices were introduced. In Lucknow, social control was imposed on urban design, with new building regulations imposing uniform and regular architecture, suppressing balconies and 'ornate' design. In the new colonial suburbs, traditional courtyard house-forms supporting segregation of sexes and the practice of purdah, were considered by the new rulers to be 'ill-designed' and 'poorly ventilated' and were not allowed; if Indians moved there, their bungalows had to conform to Western designs. The measures marked 'a steady revolution in the style of elite housing in the city' (Oldenburg, 1984; King, 1976; 1984); elsewhere in India, 'Western' forms were introduced with new tastes expressed in architecture and material consumption (Bayly, 1983). In Algiers, the forces of French colonialism demolished indigenous quarters and mosques, replacing them with the 'Beaux Artes' forms of the 'modern' city (Wright, 1987).

These, the 'urban rubrics' of the metropolitan cities are used to transform colonial cities in the second half of the nineteenth century. In the early twentieth century, new and more subtle policies are adopted with the new urbanistic practices of the planning profession (see Rabinow, 1989a; also Chapter 2, this volume).

State forms of government and legislation, developed in Britain and transplanted to India, are used, along with land policies developed in Australia, by the French to administer North African cities; 'tropical' architecture and medicine (for Europeans in the colonies) develops; international exhibitions — on trade, health, the colonies, planning, manufactures, colonial products, take place in London, Paris, Chicago, Vienna; international agreements are concluded and international associations formed. 'International', of course, is to be understood in this context to include only the capitalist nations of the industrialized west. These activities increasingly develop in the twentieth century with the first 'international' conference devoted to questions of colonial urbanism in Paris in 1931 (Betts, n.d; *Lotus International*, 1980; Wright, 1987).

An overriding concern of the administrators of the nineteenth-century metropolitan city, and increasingly of that in the colonies, was one of health. Earlier, relatively high rates of morbidity had been tolerated as the price to be paid for economic returns. But with the rapid growth of medical science in the nineteenth century, colonists' expectations about their own health, and expected lower rates of mortality, rose. Two aspects of this situation can be mentioned, both of which had implications for the form of colonial urbanism and urbanization.

The first concerns the nature of explanations for disease (cholera, typhoid, and malaria) and their essentially cultural origin, particularly, the so-called 'miasmic theory' — that disease resulted from polluted air. It was therefore believed that it could be avoided by, for example, lowering air temperatures and distancing oneself from perceived sources of infection, especially the indigenous population. From these ideas came the low-density layouts, spacious suburbs, and separation from 'the native city' (a similar point is made in Blussé's (1985) discussion of the ecology of Batavia). Ross's theory concerning Africans as the carriers of malaria and of mosquitos as the agents of transmission was to result in an official policy of segregation (Frenkel and Western, 1988). Similar beliefs (as well as other factors) help to explain the hill station, one of the most widespread manifestations of colonial settlement in South and South-East Asia. As Reed's (1976) study of Baguio in the Filipinos shows, the hill station depended, firstly, on the particular cultural perceptions and social activities of the colonial population, but more particularly, on the political power that enabled these perceptions and expectations to be met. In the early years of this century, despite protests from the Filipinos, over four million dollars were spent simply to construct the road leading up to the site of Baguio. Today, with a summer population of half a million, it is one of the most flourishing resorts in South-East Asia (Reed, 1976; see also Spencer and Thomas, 1948).

In regard to city building, 'modern' concepts of design, relying on capital-intensive technology and materials, were introduced; significantly, reinforced concrete was brought into Casablanca (by August Perret in 1914) to build dock warehouses. Here and elsewhere, traditional skills and materials were replaced, new occupations were created, and a myriad ramifications occurred as the colonial cities became economically, technologically, and culturally dependent on changes in the European metropolis (Betts, n.d; though other colonial administrators attempted to address this problem. See Rabinow, 1989a; Irving, 1983). The transfer of this transnational urban culture is facilitated by the international flow of capital.

In one sense, the apparatus of the modern city was to have similar effects wherever it was set up: a printing press, a railway station, an electric tramcar, or an automobile operate virtually in the same way, irrespective of location. Yet how they are used, who and what for, depends not only on the distribution of economic and social power and cultural factors but on the political economy of the State (including colonialism); aircraft can take some people to sunbathe on the Mediterranean and others, on a religious pilgrimage to Mecca. In the colonial context, the apparatus of the 'modern' city was first introduced to benefit the colonial (not colonized) population.

In some places, the colonies were seen as virgin fields on which the new profession of architect-planners could fulfil their imperial dreams: Lutyens in Delhi, Henri Prost in Morocco, Burnham in the Philippines, Baker in Pretoria, Adshead in Lusaka, Griffin in Canberra, Ernest Hébrard in Indo-China, and in Algiers, Corbusier, who produced seven plans for the city between 1931 and 1942 (Betts, n.d.). This was the era of the 'dual city', whether in Dakar, Algiers, Cape Town, Delhi, Nairobi, or Singapore. 'Between European Dakar and native Dakar', wrote the architect, Toussaint, 'we will establish an immense curtain composed of a great park' (Betts, 1969). These were the forerunners of Corbusier's Chandigarh, and Doxiades' Islamabad.

Another part of this global urban culture was the reflective consciousness of change, the notion that one form of civilization was replacing another. In metropolitan societies it was experienced in the late nineteenth century as an intellectual and emotional reaction to industrial urbanization, and even capitalism itself; it had its practical expression in a pseudo 'anti-urbanism' of which one dimension was the preservation of past buildings and environments (in England, supported by such influential figures as William Morris and John Ruskin); another was the idea of 'the garden city'. It was these ideologies, as well as the embryonic phenomenon of tourism (its roots in the growing system of unequal exchange) which, in the twentieth century, account for the introduction of 'preservationist' policies in the colonies: the activities of Curzon in India or in Morocco, the decision of the Résident Général, Lyautey, to leave the traditional city untouched. Elsewhere, it led to the ideas of the garden city being introduced into colonial planning.

In any comparative study of colonial cities, a key issue to explore is that of land policy. In the introduction of planned towns in many areas, the basic assumption of a free market in land, central to the growth of the capitalist city in the West, was introduced. Abu-Lughod (1975: 98–9) demonstrates for Moroccan cities how sophisticated tenure systems, connected to social and legal obligations, were simply swept away and replaced by a free market in land. Elsewhere, the colonial state became the main landowner, bequeathing to a future independent nation the problems and advantages of government control. Clearly, the building of the colonial city, from the seventeenth to the twentieth centuries, created, either for colonial landlords or others, surplus values, whether in speculative development or as rent. With the commodification of land and the development of wage labour, surplus wealth is invested by an increasingly 'Westernized' elite and middle class on housing and on consumer goods to fill them. The city becomes a unit for consumption as well as for production and trade.

CONCLUSION

The problems in explaining the colonial city are to identify the changes that result from colonialism (a power relationship of dominance enforced by an alien culture), from capitalism (a particular mode of production), from industrialization (a system of technology and energy base, and including changes in economic and social relations), and 'Westernization' (a set of cultural institutions and values, themselves differentiated according to various colonial powers — Dutch, French, British, Portuguese, etc., as well as being differentially socially expressed). Whilst all these processes interacted in the colonial city, there are none the less other examples of non-colonial 'modernizations' that provide a point of comparison (see Abu-Lughod, 1971).

It might, in conclusion, be worth acknowledging the European ethnocentricism and 'backward-looking' historical bias of this account. By the nature of one's limited knowledge, it is understandable (if inexcusable) that the colonial city is seen from 'outside'. Karasch (1985) suggests that it was not Brazil that made Rio but Rio that made Brazil. How true is it to suggest that in 'the West' the society gave rise to the city but in colonial societies, it was the city that has made the 'modern' society? In the countries where such ex-colonial cities exist, they will, as de Bruyne (1985) points out, be perceived in a totally different light. Identifying the 'colonial' in the city of 'the other' may be another version of 'Orientalism' (Said, 1978). What my unilateral view has underplayed is the contribution, within an increasingly pervasive global capitalism, of the indigenous society and culture. The next book on 'colonial' or 'ex-colonial cities' might come from representatives of those cities themselves.

Chapter Three

INCORPORATING THE PERIPHERY (2)

Urban planning in the colonies

We want not only England but all parts of the Empire to be covered with Garden Cities.

(Garden City, 2 (15) 1907)

I hope that in the new Delhi we shall be able to show how those ideas which Mr. Howard put forward . . . can be brought in to assist this first Capital created in our time. The fact is that no new city or town should be permissible in these days to which the word 'Garden' cannot be rightly applied. The old congestion has, I hope, been doomed forever.
(Swinton, G., Capt. (Chairman, London County Council, Member of the Planning Committee for New Delhi) (1912) 'Planning an Imperial Capital', *Garden City and Town Planning*, 2 (5) (NS))

In 1945, the sphere of influence of the English (Town and Country Planning) 1932 Act as developed in the West Indies extended to the continent of Africa . . . Uganda, in . . . 1948, adopted the original Trinidad scheme. . . . The influence spread further afield. Fiji legislation shows distinct traces of its West Indian origin . . . Aden has town planning laws . . . closely related to the Uganda ordinances; Sarawak has legislation which finds its roots in Sierra Leone and Nyasaland.

In the Indian Ocean, the Seychelles passed legislation derivative from the English Act. Mauritius . . . adopted legislation based on Uganda.

Thus, the 1932 Act has left its mark in all corners of the world. . . . Modern developments in British planning procedure

44

are followed closely by colonial planning officers who are always ready to profit by the experience of the mother country.

(P.H.M. Stevens (1955) 'Planning Legislation in the Colonies', *Town and Country Planning*, March)

The (town) council's staff have no formal or urban design skills and the entrepreneurial approach to urban management which has become familiar within British local authorities is not practised here. I am therefore hoping to inject an element of this approach in Nausori. . . .

(J. Gardener (Planning Assistant, Gloucester City Council and Volunteer, Voluntary Services Overseas, Department of Town and Country Planning, Suva, Fiji) (1987) 'Town planning in Fiji', *The Planner*, September: 26)

PLANNING AND IMPERIALISM

The last fifteen or twenty years have seen the growth of a new academic specialization, the history of urban and regional planning. Until the early 1980s, the focus of these studies had principally been on the development of urban planning in the core countries of Europe and North America but in recent years, there has been increasing interest in planning history in the Third World (see *Third World Planning Review*, 1979–present)

These are not two spheres of operation, however, but one. As set out in the previous pages, urban-industrial development in the core depended on the materials and markets in the periphery just as peripheral cities depended on the injection of capital and 'professional' expertise from the core. Colonial economies played a role in the industrial urbanization of Europe, especially Britain, and hence, indirectly as well as directly, in the development of 'modern' urban planning. With regard to, for example, two early influential examples of planning, where did Bournville's cocoa come from, or, in relation to Port Sunlight, the coconut oil for Lever's soaps? This chapter, therefore, examines urban planning in the colonies and some of the issues that research in this area needs to raise.

Conceivably, the scope of such a theme would cover the activities of those metropolitan societies — France, Britain, Belgium, Portugal, Spain, the Netherlands, Italy, Germany, the United States, and also South Africa, with colonial possessions in South and South-

East Asia, middle America, and Africa over the last hundred or more years. (In Africa in 1919 this included Algeria, Tunisia, Morocco, French West Africa, French Equatorial Africa, Somaliland, Togoland, Cameroon, Madagascar, Congo, Guinea, Angola, Mozambique, Cabinda, Rio de Oro, Rio Muni, Spanish Guinea, Eritrea, Italian Somaliland, Gambia, Sierra Leone, the Gold Coast, Nigeria, South-West Africa, Bechuanaland, Swaziland, Basutoland, Southern Rhodesia, Northern Rhodesia, Nyasaland, Tanganyika, Zanzibar, Uganda, Kenya, the Sudan, and Egypt.) To be technically correct, it would be necessary to include other continents — such as Australasia — politically defined as 'colonies' during part of the period. It should purportedly deal, not only with the 'grand designs' for Delhi, Lusaka, Canberra, Salisbury, Nairobi, Kaduna, or Kuala Lumpur, but with a myriad cases from Fez to Djibouti, Casablanca to Luanda, where some forms of conscious planning took place, if not in the particular British manner of town-planning ordinances, improvement trusts, or master plans (Alcock and Richards; 1953; Atkinson, 1953). It would deal with the activities of Lutyens, Baker, Geddes, and Lanchester in India (Delhi, Calcutta, Madras), with Baker or White in South and East Africa (Pretoria, Nairobi), with Reade in Malaya, Adshead in Northern Rhodesia (now Zambia), Ashbee in Palestine, Gardner-Medwin in the West Indies, or their American, Italian, French, Portuguese, Dutch, German, Belgian, or other counterparts in the Philippines, Morocco, Algeria, other parts of colonial Africa, in Saigon, Cholon, or the Treaty Ports of China or other parts of the colonial urban world (see Abu-Lughod, 1975; 1976; 1978; 1980; Boralévi, 1980; Christopher, 1988; Davies, 1969; Dethier, 1973; *Garden City*, 1904–10; *Garden Cities and Town Planning*, 1911–32; *Town and Country Planning*, 1932–80; Gardner-Medwin, *et al.*, 1948; Fetter, 1976; Ginsburg, 1965; Hines, 1972; Home, 1983; Langlands, 1969; Lewcock, 1963; Rabinow, 1989a,b; Reitani, 1980; Sandercock, 1975; Shapiro, 1973; Simon, 1986; Soja and Weaver, 1976; Wright, 1987; 1989; Wright and Rabinow, 1981; Western, 1985). If this is the possible scope for the field, this chapter is limited to the British colonial experience as it relates to those areas where subject populations were incorporated on a large scale into the political, economic, social, and cultural systems of the metropolitan society.

Two conceptual clarifications are needed. As indicated in Chapter 2, colonialism is understood as 'the establishment and maintenance, for an extended time, of rule over an alien people that is is separate from and subordinate to the ruling power' (Emerson, 1968). In this case, we are referring to 'modern' industrial colonialism, mainly of the nineteenth and twentieth centuries. For a proper understanding of the processes of 'colonial planning', some basic conceptual distinctions would need to be made between different colonial situations, for example, colonies of settlement, and those of exploitation of the indigenous inhabitants, as also the number of indigenous people in a colonized territory, their state of economic development, and the length of colonial rule (see Chapter 1; also Christopher, 1988; Drakakis-Smith, 1987).

Second, what is to be understood by 'urban and regional planning'? The history of urban planning in any society demonstrates a continuity, in terms of the distribution of political and economic power, and in social and cultural values, between an age when there was no 'town planning' as such, and a period (after 1909 in Britain) when there was. This continuity in practice is especially evident in colonial territories (King, 1976) where the urban assumptions of industrial capitalism combine with cultural practice to produce particular spatial forms well before a 'professional' expertise of 'town planning' is constructed. In this context, therefore, the modern history of colonial planning can be usefully divided into three phases:

1. A period up to the early twentieth century when settlements, camps, towns, and cities were consciously laid out according to various military, technical, political, and cultural codes and principles, the most important of the latter being military and political dominance. These practices are the outcome of the state of development in the core society (economic, social, and political), the mode of production on which this was based and the (often military) context in which they had developed.[1]
2. A second period, beginning in the early twentieth century that coincides with the development of formally stated 'town-planning' theory, ideology, legislation, and professional knowledges in Britain, when the structure of colonial relationships was used to convey such phenomena, on a selective and uneven basis, to the dependent territories.
3. A third period of post-, or neocolonial developments, after 1947 in Asia and 1951 in Africa,[2] when cultural, political, and economic

links have, within a larger network of global communications and a situation of economic dependence, provided the means to continue the transplantation of ideologies, values, and planning models, generally in the 'neocolonial modernisation' of once-colonial cities (Steinberg, 1984).

This periodization can be matched with classical, monopoly (or imperialist), and multinational phases in the development of capitalism and classical, 'modern', and most recently, 'post-modern' phases in the development of culture, particularly as it relates to architecture and planning (see comments on Jameson, 1985: 11–12). This is discussed in more detail on p. 8.

Moreover, in discussing urban planning in relation to colonial territories, it is impossible to disassociate a more limited notion of 'planning' from, at one level, a range of related topics such as architectural style, health, house form, legislation, building science, and technology and these, at another level, from the total cultural, economic, political, and social system of which they are a part. The introduction of 'modern' 'planned' environments based on 'Western' (and capitalist) notions of civilization, when compared to the 'traditional' (non-capitalist) indigenous city of Kano (northern Nigeria) or the Malayan village, has obviously modifed far more than just the physical environment.

The aim of this chapter, therefore, is to suggest five interrelated themes that are central to any discussion regarding urban planning in a colonial and neocolonial situation.

THE POLITICAL-ECONOMIC FRAMEWORK

Colonialism, as a political, economic, and cultural process, was the vehicle by which urban planning was exported to many non-Western societies. How were the aims and activities of planning affected by this? How did urban and regional planning contribute to or modify the larger colonial situation? What is the present-day structure of political and economic relationships that enables the assumptions and aims of planning to be transferred to ex-colonial societies?

A consideration of these questions presupposes an intellectual and moral stance. Stated as a simplistic dichotomy, this may be an analysis that treats imperialism as 'the highest state of capitalism' (Lenin) where subject populations (and environments) were incorporated into the metropolitan capitalist economy with the attendant consequences for good or ill inherent in that system; at the other end of the spectrum, colonialism can be seen as the primary channel by which the benefits of Western civilization, 'the ideas and techniques, the spiritual and material forces of the West' have been brought to a large portion of humankind (Emerson, 1968), a viewpoint shared by some African scholars whether Black or White (Adu Boahen, 1987; Christopher, 1988). There are also, of course, positions in between.

In the simplest analysis, colonialism was a means by which the metropolitan power extended its markets for manufactured goods and by which the colonies, in turn, supplied raw materials to the metropolis. Though this oversimplifies the historical situation, economic dependency characterizes both colonial and post-colonial situations, with the concomitant phenomenon of what Castells (1977) has termed 'dependent urbanization', i.e. industrialization that historically was related to urbanization in the development of core Western societies, which, in the colonial case, took place in the metropolitan society, whilst urbanization (without industrialization) took place in the dependent colonial society. Urban planning in the colonies, along with its associated activities such as housing and transport developments, in any of the three periods indicated, may be viewed as part of 'dependent urbanization'. The evolution of urban systems (regional planning) and the organization of urban space (urban planning) in the colonial society can be accounted for by 'the internal and (especially) the external distribution of power' (Friedmann and Wulff, 1976: 13–14). Despite criticisms of the dependency paradigm (see Corbridge, 1986) Friedmann and Wulff's comments on it still have salience for the understanding of historic colonial environments:

> Basically, it involves the notion that powerful corporate and national interests, representing capitalist society at its most advanced, established outposts in the principal cities of Third World countries, for three interrelated purposes: to extract a sizable surplus from the dependent economy, in the form of primary products, through principally a process of 'unequal exchange'; to expand the market for goods and services produced in the home countries of advanced monopoly capitalism; and to ensure stability of an indigenous political system that will resist encroachment by ideologies and social movements that threaten to undermine the basic institutions of the capitalistic system. All three forms of penetration are ultimately

intended to serve the single purpose of helping to maintain expanding levels of production and consumption in the home countries of advanced capitalism.

(Later versions of this theory see the domination of peripheral economies primarily of help in the expansion of multinational corporations that exhibit a growing independence of action from national commitments and control.)

In the course of this process, local élites are co-opted. Their life-style becomes imitatively cosmopolitan. . . . As part of this process massive transfers of rural people are made to the urban enclave economy.

(Friedmann and Wulff, 1976)

According to Friedmann and Wulff, this theory of dependency, or more accurately, dependent capitalism, seems to account for certain forms of spatial development in newly-industrializing societies (ibid).

These theoretical generalizations can be translated more specifically into colonial built forms and urban spaces. Colonial forms of urbanism and urbanization are evident not only in Latin America, Asia, and Africa (Basu, 1985; Castells, 1972; Harvey, 1973; King, 1976; Mabogunje, 1980; Ross and Telkamp, 1985) but also in the early-eighteenth-century American colonies (Foglesong, 1986; Gordon, 1984; Lampard, 1986) as well as in Australia.

In India, the most widespread example of conscious urban planning prior to the twentieth century is the location and lay-out of military cantonments, the primary purpose of which was to provide for the ultimate sanction of force over the colonized population. The informally planned 'civil station', located alongside, accommodated the political-administrative 'managers' of the colonized society. The major cities resulting from the colonial connection (Madras, Bombay, and Calcutta), were not industrial centres but commercial, entrepôt ports oriented to the metropolitan economy (Brush, 1970; Kosambi, 1985) subject to nineteenth-century planning exercises to optimize their function in the colonial economy (Dossal, 1989). The major city-building exercise during two hundred years of informal and formal colonial rule — the planning and construction of New Delhi (1911–40) — involved the creation of a capital city almost entirely devoted to administrative, political, and social functions with virtually no attempt made to plan for industrial development. The so-called 'hill stations' — a major example of specifically colonial urban development — had

50

primarily political and sociocultural, consumption-oriented functions, the most famous of them, Simla, described by Learmouth and Spate (1965) as 'parasitic' in relation to India's economy (King, 1976: 156–79).

In Africa, for much of the colonial period, the functions of newly established centres were political, administrative, and commercial. The built environment of the 'ideal-type', political, administrative capital was characterized by those buildings housing the key institutions of colonialism: the government or state house, the council or assembly buildings (if any), the army barracks or cantonment, the police lines, the hospital, the jail, the government offices, and the road system, housing, and recreational space for the ex-patriate European bureaucracy, and occasionally, housing for local-government employees. At its most extreme, as at Lagos, the place of the central business district was occupied by the race course.

The economic institutions of colonialism were expressed in physical and spatial form: the penetration of finance capital in the construction of banks, insurance buildings, and the headquarters of multinational corporations; the incorporation of labour power in the 'native townships', mining compounds and the 'housing for labour'. The whole relationship between the economic system, the supply of 'native labour' by induced migration from rural areas, and the provision or, more accurately, lack of provision for their accommodation is too large an issue to be discussed here, 'having', to quote Collins (1980), 'a literature of its own' (see also Amin, 1974; Gugler, 1970; Rex, 1973). The underlying assumption of the system of circulatory labour migration in the Copperbelt was that:

> the towns were for Europeans and the rural areas were for Africans. It followed that no African should be in town except to provide labour as and when required by a European employer. Only men were required. . . . Urban housing was therefore rudimentary.
> (Collins, 1980: 232)

In the pre-independence phase in Africa, considerable effort was placed in planning and construction of low-cost and 'planned' 'African housing'. Such efforts can be viewed on the one hand as evidence of changing values and priorities, concern for what were seen as the unsatisfactory conditions of indigenous rural migrants living in shanty towns on the edge of the city: 'planned' housing, with sanitation, electricity, and

51

a water supply constructed as part of a new social welfare programme of the Colonial Office. It can also, however, be seen as a means of incorporating labour into the colonial economy. Thus, 'social housing' built in the Gold Coast in the 1950s is 'to house labour required at the harbour'; housing in Jinja, Uganda, are 'units for Labourers, Waluka Labour Estate'. Housing in Nairobi in the late 1920s, and elsewhere, was constructed on the 'bed space' principle to accommodate single male labourers *(Colonial Building Notes (passim))*. Built and let by the metropolitan government, housing in this sense represented a subsidy to wages — either to government employees or metropolitan enterprises — as well as a potential instrument of social control. The planning, design, and building of such housing estates — as also of 'Asian' and 'European' housing (of a somewhat different order) — in separate parts of the town is patently 'town planning' and the nature of the activities are part of the larger colonial enterprise.

For example, according to the Report on the Rhodesian African Home Ownership Scheme, cited in *Colonial Building Notes*:

Between 1945 and 1957, the number of Africans employed in urban factories in Southern Rhodesia (now Zimbabwe) rose from 95,000 to 300,000. 'This increasing demand for labour was most fortunate for whilst it was developing, the population of the rural areas was rising rapidly, largely because of the steep reduction in infant mortality which followed the introduction of medical services into the countryside'. In 1953, the shortfall of 'single quarters' in Salisbury was 13,500. As 'a certain amount of disquiet had been felt about the earlier system of housing large numbers of single men in hostels' it was thought that such in-migrants would be more comfortable and 'socially more stable' if married householders could be encouraged to accept lodgers of their own choosing (each working African receiving £1 housing allowance each month). In the housing scheme devised, costs were cut to a minimum to result in over 2000 units of housing being built for approximately £140 each outside Salisbury (now Harare). The Treasury Accounts report, 'The cost per square foot obtained is probably the lowest on record in Africa and it is possible that a world low record has been achieved for the construction of this type of building today'.

(Colonial Building Notes, 1959, no. 60, June)

The metropolitan government, in developing these low-cost housing programmes, generates and exchanges information on standards, costs, and design with other European powers with interests in Africa: the Belgian Office des Cités Africaines in the Congo (later Zaïre); the French Bureau Central d'Etudes pour les Equipments d'Outre-Mer (on 'tropical housing') of the Secrétariat des Missions d'Urbanisme et d'Habitat; and the South African National Housing and Planning Commission (on minimum standards for housing non-Europeans). Information is also exchanged with major metropolitan multinational companies with their own housing programmes for African workers: Imperial Tobacco, Fyffes, or the Union Minère (later, Gecamines) in the Congo (*Colonial Building Notes, passim*). In brief, 'official' housing and planning policy is primarily directed to ensuring basic minimum standards for the local labour force and government employees as well as government buildings, administrative buildings, and, for collective consumption, welfare buildings (schools, hospitals, and colleges). Industrial development, such as it is, is the responsibility of local, or more usually, metropolitan based multinationals.

THE CULTURAL, SOCIAL, AND IDEOLOGICAL CONTEXT OF COLONIAL PLANNING

The history of 'town and country planning' in Britain in the industrial-capitalist and post-industrial era is, in one sense, a unique and culture-specific historical experience. True, common factors resulting from the influence of industrialization or modern automative transport may induce a structural similarity in urban environments of different industrial and capitalist societies: in some respects, Birmingham is like Berlin in the same way as pre-industrial Fez is like pre-industrial Baghdad or Katmandu. Yet given such economic or technological influences, the extent to which urban forms and planned environments differ clearly depends on political and economic factors, cultural values, historical experience, geography, and the values and ideological beliefs of those power-holding groups and professional elites responsible for structuring and implementing decisions about urban planning and the overall shape of towns.

The particular ideological and cultural context of British planning in the first half of the twentieth century, as dominated by the 'Garden City movement', is well known. The primacy of 'health, light and air', combined with a set of social and aesthetic beliefs, as a reaction to the nineteenth-century industrial city was expressive of an implicit environmental determinism that pursued physicalist solutions to social, economic, and political ills ('the peaceful path to reform'). It was a strategy of power exercised by municipal authorities to alleviate what were defined as social pathologies.

From this nineteenth- and early-twentieth-century experience grew the theory of physical planning, as well as planning legislation and the mechanisms to implement it, a form of social technology in which environments, and people, were modelled or controlled in accordance with an assumed 'public good'. It was this 'expertise' that, with its assumptions, values, and practices and partly modified by local conditions, was exported to colonial societies. There are many aspects of this process only some of which can be touched on here.

As discussed in the previous section (pp. 48–53), physical-planning notions and legislation were introduced as part of the overall situation of colonial power. The basic divisions of the society, political, social, and racial, inherent in the colonial process, between ruler and ruled, Black (Brown) and White, rich and poor, 'European' and 'native', were taken as givens. In this situation, the 'techniques' and goals of planning — 'orderly' development, easing traffic flows, physically 'healthy' environments, planned residential areas, reduced densities, and zoning of industrial and residential zones were introduced, each according to the standards deemed appropriate to the various segregated populations in the city — and all without disturbing the overall structure of power.

Second, the overriding, even obsessive concern with 'health' (referred to by Swanson (1970) as the 'sanitation syndrome') was, after the implicit political and economic function of planning, taken as the driving force behind planning in all colonial territories. The creation of physically 'healthy' environments, defined according to the cultural criteria of the metropolitan power, became a major objective. It is 'health' rather than health because the basically relative nature of health states and their overall cultural and behavioural context (discussed on p. 55), if appreciated at some times, were ignored at others. Indigenous definitions of health states, the means for achieving them, and the environments in which they existed were replaced by those of the incoming power in a total ecological transformation.

Thus, vital statistics from the metropolitan society are used as the reference point to 'measure' health states in the colonial population; historically and socially derived concepts of 'overcrowding' developed in the metropolitan society are applied to the indigenous environment and people. In the interests of 'health' and the new economic and social order, new environments are created — rows of minimal 'detached' housing units, surrounded by 'light and air', 'open space', gardens, and recreational areas in total disregard of the religious, social, symbolic, or political meaning of built environments as expressed in the indigenous villages and towns. It is instructive to compare the evaluation of African housing environments in Paul Oliver's (1971) *Shelter in Africa*, or his more recent (1987) *Dwellings. The House Across the World*, or Susan Denyer (1978), *African Traditional Architecture*, with the following extract from the *Annual Report on Medical Services*, 1953–4, Federation of Nigeria, 1955:

> the time-honoured, mud-walled compound with its intricate rabbit warren of ill-lit, ill-ventilated and undersized rooms. The more wealthy spend much on beautifying the front of their houses . . . though housing is bad enough in the large towns it becomes more and more primitive until one reaches the encampments of the nomadic cattle Fulani where large families exist in small wigwam shelters made of grass matting. . . . During the year a housing survey was made in Argunga town . . . to assess what degree of overcrowding existed. The standards of calculation used were those of the *England and Wales Housing Act*. The results were surprising in that in most of the town there was little or no overcrowding where cubic space per person was concerned.
>
> (*Colonial Building Notes* (1956), no. 41, December)

Health care defined according to metropolitan cultural norms, with its systems of inspections, regulated environments, and controls over behaviour becomes, like the police, housing, or employment, another means of discipline and social control. Because of the racially segregated nature of the society, as Swanson (1970) points out, 'problems of public health and sanitation, overcrowding, slums, public order and security are perceived in terms of racial differences'. Though many of the objectives of municipal government (abatement of health dangers, slum clearance and housing) were legitimate, in a colonial society the pursuit of class interest and the exercise of prejudice regarding race, culture, and colour mixed up these objectives with racial and social issues (ibid). The culture and class-specific *perception* of health hazards more than the actual health hazards themselves was instrumental in determining much colonial, urban-planning policy.

From a purely physical and spatial viewpoint, environmental

standards, norms of building and design (as well as the urban institutions themselves) derived from the historical experience of the capitalist industrial State and overlaid with its particular cultural preferences, were transferred to societies with totally different economic and cultural experience (United Nations, 1971; Mabogunje *et al.*, 1978).

Where substantial numbers of 'ex-patriates' or 'settlers' were involved as in Lusaka, Nairobi, or Delhi, very low-density residential developments were built to suit their convenience. In the Master Plan for Nairobi of 1948, revisions to the original lay-out of the town, founded in 1896, suggested that densities in the European area be raised from 1 to 15 per acre (White *et al.*, 1948). Low densities, extensive intra-urban distances, large housing plots, and lavish recreational space were all based on the assumption of the availability of motorized transport and the telephone, as well as cheap 'native labour'; i.e. on a technology for which the colonized country was dependent on the metropolitan. The assumption in such plans was presumably that the 'industrial', fully motorized society was inevitable, an assumption which, in the post-colonial era of independent development has meant not only vast journeys to work, but excessive expenditure on basic services (water, sewers, roads, electricity), inefficient land use and a need for fundamental redensification.

As metropolitan environments were introduced, or rather, colonial versions of such environments,[3] so metropolitan legislation was necessary to maintain them; hence, the widespread introduction of the 1932 Act and other legislative codes (see opening quotations, p. 44). Here, two points can be mentioned. The first concerns the transfer of particular social and environmental categories from the metropolitan to the colonial society, of which the basic dichotomy between 'town' and 'country' was one of the more important.[4]

Another is the transfer of particular culture-specific practices, and especially, those values of historicism and sentiment expressed in the 'preservation' syndrome. In the colonial context, this has a double irony. Not only does planning effort go into inculcating the colonized culture with similar values but the criteria of the colonial power are used to define and 'preserve' 'buildings of architectural and historic importance', while remnants of the indigenous culture are allowed to disappear. Thus, the Ministry of Overseas Development-sponsored Survey and Plan for Kaduna, Northern Nigeria, 1967 (Max Lock and Partners, 1967) suggests the retention of a small iron bridge erected by Lord Lugard, the previous colonial Governor; the Secretary of the

Georgian Group visits the West Indies to advise on the preservation of military officers' quarters from the eighteenth century. In Delhi, various 'sacred' sites, associated with the 'Mutiny' are preserved throughout colonial rule, indirectly affecting the location of the new capital (King, 1976: 234).

COLONIAL PLANNING: SOCIAL SPACE

The central social fact of colonial planning was segregation, principally, though not only, on racial lines. The segregated city not only resulted from but in many cases, created the segregated society. In southern Africa, the indigenous population was kept out of cities; here and elsewhere it was confined to 'native locations' or 'townships' (Soweto, of course, stands for the South West Township), or it 'squatted' on the perimeter. In India, an implicit apartheid based on economic and cultural criteria governing occupation of residential areas was practised. In other south-east Asian cities zoning of Asian and European areas was the norm (McGee, 1967).

In South Africa, as labour migration increased, 'native housing' was provided in locations on the edge of the city. As urbanization proceeded, Africans were 'brought into' the urban systems in the form of segregated cities, thus, as Swanson (1970) describes, learning to see themselves in the new social categories imposed by the ruling White minority. 'The urban nexus explains why the policies of segregation and separate development emerged as the dominant concerns of local and national government', the *Native (Urban) Areas Act* of 1923 embodying, for the first time, national recognition of the impact of urbanization. The segregated city has been fundamental in the development of 'categorical' relationships, the stereotyping of one race and its behaviour by another (Mitchell, 1966).

Even within the larger racial divisions, transformations have also occurred as a result, in later times, of particular planning and housing policies that have allocated different social groups to housing types and residential areas built and allocated according to economic (i.e. income-bracket) criteria. Because of the lack of finance or capacity in the private sector, in many colonial and ex-colonial societies, a large proportion of housing has been undertaken by government, particularly in newly created urban centres. The design and allocation of housing and area according to occupation and income group have been significant in structuring perceptions of social stratification (King,

57

1976; Little, 1974; Nilsson, 1973). Similar practices — a continuation of the colonial Public Works Department tradition — can be found in Chandigarh or Islamabad.

Nothing could be more different than the traditional Ashanti village and the low-cost, gridiron, planned, suburban housing-unit estate of Accra. The symbolic meaning of space in the traditional village, whether expressed in terms of house or compound size, dwelling form, or distance between dwellings, in all cultures relates to social, cultural, or religious meanings. New urban environments based on income and occupational differentials clearly affect both the construction and self-perception of social classes and categories. Yet in Ghana, it was assumed that such planned housing could be used as a means to break down traditional tribal and kinship bonds and help to establish a 'law-abiding' and, with the introduction of privately owned, single-family dwellings, an implicitly consumer society:

> As urbanization takes effect in Ghana, tribal ties and discipline must be superseded by other loyalities if a coordinated law-abiding society is to emerge. It is therefore important to give the urban Ghanaian a sense of community membership. The policy in Tema has been to discourage racial, tribal, religious or class segregation, in the hope that the citizens' loyalty will be to neighbourhood, community and town. This policy requires non-traditional types of housing accommodation. The tribal compound has no place in Tema and is replaced by the private family dwelling. Differentiation of dwelling standards is purely by income and all income groups are represented in each community.
>
> (*Tema, 1951–61, A Report on the Development of the Town of Tema*, prepared for the Ghana Government by D.C. Robinson and R.J. Anderson)

(In this case, Russian town planners had also submitted a proposal to the Tema Development Corporation for the development of one 'community area' and based on a more physically collective conception of community living. This included high flats, communal kitchens, etc., which, at an estimated cost of £8m, was much higher than that of other 'community developments'. However, 'the scheme was not considered at all suitable to the Ghana way of life and it has not been accepted' (ibid).)

THE INTERACTION OF ENVIRONMENT AND BEHAVIOUR

Urban planning relates, on the one hand to the actual creation of planned space and, on the other, to the regulation and modification of existing areas by means of statutory legislation and municipal controls. In democratic societies, it is assumed that statutory control — the law — represents the 'collective will' of society, the contested outcome of economic and political interests, but which also has the power to change it. In theory, therefore, members of the society are, by and large, in agreement with the law and, in a stable polity, accept it as legitimate.

A more important factor controlling the use and modification of the environment — determining how houses are built, how public space is used, how people behave in specific areas — is the whole realm of 'unwritten law', the taken-for-granted rules and codes, based on shared values that are part of everyday cultural practices and behaviour.

In the case of planned environments as well as planning legislation exported to culturally different, pre-capitalist societies, neither of these two assumptions applies. By definition, such societies are not democratically governed. Legislation is imposed after being conceived for the interests of the ruling elite. In ensuring that such legislation is enforced, resort must be had to the instruments of such control — the police, the army, and the judiciary, or the informal, but effective, para-judicial policing by members of the ex-patriate community.

In a totally different culture, the taken-for-granted codes and cultural rules governing people's relation to their environment simply do not apply to culturally different 'imposed' environments; indigenous codes conflict with those of the newcomers, most obviously, in pre-capitalist, pre-industrial situations, with the entire building process and the way that space is organized and used. Hence, over time, two interrelated processes take place. New laws and regulations are enforced by municipal or State authorities by a mixture of penalty and example; second, the life style and cultural behaviour of local populations may be modified as they emulate the ruling colonial elite. These regulatory mechanisms are buttressed by the power of the State: in cases, keeping indigenous populations out of cities and/or distributing resources in favour of colonial populations.

COLONIAL AND NEOCOLONIAL PLANNING:
THE CONTENT AND MECHANISMS

The ideological content of planning and the mechanisms by which it was transferred between the metropolitan core and the colonial periphery (with regard to Britain) are best understood within the periodization set out at the beginning of this chapter.

The first period, covering the late eighteenth and nineteenth centuries, is one of classical industrial capitalism and prior to the development of 'professionalized' 'Town Planning'. None the less, planning takes place, its objects (the cities of the colonial system) being seen as extensions of the metropolitan space economy. In colonies of settlement, urban plans and forms based on inherent economic, legal, and social principles, are transplanted — with some local modifications — from the metropole: concepts of property, the notion of a land market, prevailing levels of technology and transportation, cultural and social assumptions about the use of space, etc. Both the urban hierarchy that develops as well as the urban structure and architectural form, however, are equally dependent on the role of the colony in its principal task of producing raw materials for the metropole, acting as a market for manufactured goods and more generally, on its place in the prevailing international division of labour (for the example of Australia, see Mullins, 1981; King, 1984).

In colonies of exploitation, similar principles apply: the cities are 'extensions' of the metropolitan space economy. Urban-planning ideas and forms reproduce, in their particular manifestation, the political, economic, social, and cultural assumptions of the metropole, informed additionally, by the urbanistic 'knowledges' generated by the processes of capitalistic industrialization. These account for the 'classical' phase of colonial urban development. Colonial architecture is largely a reproduction of metropolitan forms, though adapted to meet the climatic, resource, and other specific needs of the colonial 'third culture' and colonial situation (King, 1976; Metcalfe, 1989). Where older colonial settlements are restructured, e.g. Bombay in the 1860s (Dossal, 1989), Lucknow in the same period (Oldenburg, 1984) or in West Africa, Accra, Freetown, or the remodelling of Lugard in Nigeria (Acquah, 1958; Frenkel and Western, 1988; King, 1984), the principles of planning are those based first, on colonial dominance, and also on the use of economic, technological, scientific, medical, and legal knowledge imported from the metropole. The phase is comparable to the period of State regulation through 'by-law' planning, building regulation, sanitary inspection, and more centralized control at the core.

The mechanisms and agencies for the transfer of these practices during this period are often military engineers; in India, the responsibility of the Military Engineering Board. Military Engineers were educated at academies at Chatham and Woolwich and, in India from mid-century, at Thomason Engineering College in Punjab (later, Roorkee University) and Madras Engineering College. Later, with the establishment of a Public Works Department (1854), surveyors and civil engineers were attached to municipalities. With the growth of the 'professionalization' of architecture, surveying, and civil engineering in the metropole from mid-century, an increasing interest was expressed in the colonies in the technical press and the first 'official architect' was appointed to the Government of India in 1858. In the late nineteenth century, an institution for the specialized training of civil engineers was established in the metropole at Coopers Hill, Surrey. Also during this period, the long-established colonial practices of building and design become articulated and formalized into a new branch of metropolitan and colonial knowledge, 'tropical architecture' (Smith, 1869). In the early years of the twentieth century, many of the principles of 'tropical architecture and planning' are transplanted, by the movement of military and government personnel, from India to Africa (King, 1984: 207–8). French colonial planning practices in Algeria seem likely to have followed on the same broad principles (Wright, 1987).

The second period, the main phase of monopoly capitalism and imperialism, runs from the early twentieth century to 1940, and in certain respects, to the mid-1960s. This includes the period of 'high imperialism' prior to 1914 and coincides with the full articulation and development, in the metropole, of 'modern' town planning and architecture and, from 1931, the so-called 'International Style' in architecture and urban design (Relph, 1987; see also Chapter 4).

In the metropole, the still-dominant position of the core states in the world-economy and the continuing imperial connection were critical factors permitting the increasing involvement of the State in the production of social housing and urban redevelopment ('slum clearance'); in this 'interwar' period, some four million new houses were built, a quarter of them with State subsidies. 'Planned' extensions to cities and remodelling of towns took place according to the new ideologies that developed, especially in relation to the experience of the first phase of capitalist urban development. In this period, the 'Garden City movement' was founded, the first 'town planning' legislation passed by the State (1909), a 'Town Planning Institute' established the full 'professionalization' of a new 'science' and 'town planning' was established as a university 'discipline'. During these years, and especially during the post-war boom of the 1950s–60s, with major legislative planning measures passed in 1932 and 1947, planning ideology and practice rested largely on the (generally unquestioned) assumptions of an urban industrial economy of a society still perceiving itself to be well-placed in terms of global economic competition. It is during this period (the early 1900s to the 1960s) that the theorized professional practice became increasingly institutionalized and accepted, to play a significant role in the major period of urban restructuring in the 1960s: this included the implementation of a wide range of 'professional' beliefs and practices, including development controls, strict control over 'green belt' development, and divisions between 'town' and 'country', the destruction and reconstitution of downtown areas with the rehousing of subject populations into high-rise 'point' and 'slab blocks'. In so far as this can be stated in a historical context, this period represents the triumph of professionalism.

This second phase also, and especially, represented the heyday of colonial planning with the situation of imperial power permitting the ideology and practices of core imperial states to be exercised in the peripheral colonies. New colonial capitals and cities included New Delhi, Canberra, Pretoria, Lusaka, Kaduna, Rangoon, as well as the continued development and 'improvement' of numerous small towns, hill stations, and military (cantonment) locations in India, Africa, and South-East Asia during this period. Likewise, for the French in Rabat, Casablanca, Saigon, Hanoi, Tananaruve (Wright, 1987), for the Italians in Tripoli, and Ethiopia (Reitani, 1980; Boralévi, 1980), the Portuguese in Angola and Mozambique, or the Germans in Windhoek (Simon, 1986) this period became a major phase of expansion and colonial planning, only comparable in importance to that of Iberian colonial urban development in the Americas some three centuries before.

In the two major imperial realms (of France and Britain), the colonial situation permitted the expression of planning ideologies that political, economic, and also spatial constraints denied their practitioners 'at home': 'There', in the words of the French Minister of

Colonies in 1945, 'space is free and cities can be constructed according to principles of reason and beauty' (Betts, n.d.). French colonial administrators, denied the possibility of introducing new urbanistic practices in France, and impatient of the delays of parliamentary government, chose to put them into effect in Morocco. Rabat-Sale and Casablanca were the results (Abu-Lughod, 1980; Rabinow, 1989ab; Wright, 1987; 1991). For the French, in this self-reflexive era of 'the modern', it was to use planning to preserve the social hierarchies and old Medina of Morocco. For the British, 'rational planning' meant the 'Garden City' (see opening quotes, p. 44). For the Italians, North Africa offered opportunities for the implementation of fascist ideologies (Reitani, 1980; Boralevi, 1980).

For the British during this period, the mechanisms were the new professional agents sent to, or appointed by colonial governments, the transplantation of legislation (see opening quotations, p. 44); the growth in and circulation of 'professional' publications, or the visits of metropolitan consultants.

The postwar period saw these links become increasingly tangible. Funded through development and welfare funds of the Colonial Office from 1940, town planning, housing, and building 'expertise' was increasingly transplanted to the colonial territories, especially in Africa, the West Indies, and elsewhere. By 1947, just under fifty British architects and planners were working in colonial administrations. In 1948, as a result of collaboration between the Colonial Office and the Building Research Station in metropolitan Britain, a Colonial Liaison Unit was set up at the BRS to deal with requests from 'overseas administrations' concerning housing, building, and planning matters and to disseminate information on these activities (Atkinson, 1953). From this arrangement emerged the *Colonial Building Notes* (1950–8), continued, when the Unit became the 'Tropical Section' and subsequently, the Overseas Division, as the *Overseas Building Notes* (1959–present). The notes, containing 'technical information' generated by the BRS and collected from colonial and other tropical areas also contain data on metropolitan and overseas planning, housing, and architectural practice and were circulated to colonial and ex-colonial societies (*Overseas Building Notes* 1971: 141; *Colonial Building Notes*, 1950–7).

By 1950, the growth of international networks of communication was so extensive as to make these localized channels less important. In particular, the existence of international organizations, especially those sponsored by the United Nations, became important means by which 'planning knowledge' was transferred to the colonial and, from 1960, to the ex-colonial world (for example, the Housing and Town Planning Section, Department of Social Affairs, the UN, ILO, UNESCO, WHO, and various national and regional organizations, see *Colonial Building Notes*, no. 32, 1955).

Compared with the small numbers of colonial students who studied in the metropolitan society prior to 1939, an increasing number arrived after 1950. The recognition that planning and architecture in these societies required distinctive skills resulted, in 1955, in the setting up — at the suggestion of the Colonial Office and others — of courses in 'Tropical Architecture' at the Architectural Association (subsequently to form the basis for the Development Planning Unit, University College, London).

From 1967, special courses in planning, now redefined in terms of meeting 'the needs and problems of developing countries',[5] were established in the metropole.

Advisory and consulting roles by firms of British planners (aid-financed by the Ministry of Overseas Development) resulted in comprehensive redevelopment plans for downtown Kingston in Jamaica, for Cyprus, for Kaduna in Nigeria, for Francistown in Botswana, and many more (Atkinson, 1953). Teaching and planning staff from 'developing countries' were sponsored to undertake courses in Britain; study tours were arranged to see examples of British planning and housing. Planning advice in the preparation of planning legislation was offered to Cyprus, Malta, Nigeria, and Trinidad and Tobago. A further link was established through professional institutes and associations, including the Commonwealth Association of Planners and the Royal Town Planning Institute (of whose over 4,000 corporate members in 1970, some 760 were 'overseas'). Since 1957, the Institute has had an Overseas Section at its annual Summer School devoted to planning in Third-World countries.

The third period, variously described as the era of multinational, global, or disorganized capitalism (Soja *et al.*, 1983; Lash and Urry, 1987; Thrift, 1986b), the new international division of labour (Frobel, *et al.*, 1980), or post-imperialism (Becker *et al.*, 1987) dates from the end of the 1960s and early 1970s. In the periphery, it followed the end of formal colonialism (1947–67) and in the core, it is marked by steady deindustrialization, a perception of loss of markets in old colonies and the need for new European replacements, a sense of the declining status of Britain in the world-economy and a growing consciousness of economic and urban crisis. From the mid-1980s, its cultural forms have increasingly been labelled 'post-modernism'. Whilst the full effects of this new phase of capitalism on urban planning in the core, and previously colonial societies, awaits full investigation (though see Lash and Urry, 1987, on economic change and spatial restructuring in the core states), the following sketches some interim comments.

In the two immediate postwar decades, physical planning was still seen as an exportable good: British experience of new towns, social housing, town-planning legislation, school design, etc., were perceived positively in their country of origin (Richards, 1961) though little, so far, is known about their reception in those of destination. From the late 1960s, however, hesitation had begun. In the ex-colonial 'developing' countries the looked-for 'take-off' (as it was still perceived in Rostow's terms) had not occurred; in metropolitan planning itself, stimulated by greater public awareness and criticism, self-doubts had set in. By 1970, planning was no longer a technical expertise but a highly politicized, value-laden activity, very much under public scrutiny. Events abroad, in Latin America, Vietnam, Africa, and the continuing poverty of ex-colonial societies, found 'development' theories of the 1960s replaced by far more radical views of the world power structure (e.g. Frank, Amin, and ECLA theories on dependency; see Chilcote, 1984). Despite shifts in the metropole to the notion of 'development planning' for previously colonial countries, increasing awareness of uneven development undermined earlier assumptions.

The energy crisis and OPEC price rises of 1973 were major factors exposing urban environments and planning to the economic and political assumptions on which they were based. Increasing economic decline and unemployment likewise exposed the assumptions on which the urban renewal schemes of the 1960s had been grounded and these themselves increasingly came under public attack. Policies of decentralization, first seen as positive achievements of planning, were later to be criticized for undermining the viability of cities. The scope (and confidence) of planning shrank, increasingly confined to small-scale interventions. Knox characterizes the radical break in urban planning of the 1970s as:

the shift away from the rationalist, functionalist, paternalist and evangelist pursuit of segregated land uses and sweeping renewal schemes towards a more participatory and activist-influenced planning aimed not only at halting renewal schemes but also at preserving and enhancing the neighbourhood lifeworld.

(Knox, 1988: 5)

These ideas were also to influence concepts of planning that were now being developed in the metropole for use in the previously colonial periphery. Whilst the changing paradigms and content of development planning are too large a topic to discuss here, as is a consideration of its mechanisms, two aspects may be mentioned briefly.

First, during the 1970s, assumptions about the relevance of metropolitan ideas and values for urban planning in 'developing countries' were increasingly questioned. From a 'liberal' perspective, there was a shift towards interest in indigenous design, the variability of standards, and the need to modify 'colonial' assumptions (e.g. Oliver, 1969; 1971; Rapoport, 1969; United Nations, 1971); though many of these issues had been raised previously, they had gone unattended (see *Habitat International*, 7, 5/6, 1983). Later, however, the larger structural relationships between core and periphery, and in the determining context in which planning took place, were increasingly exposed, first, in the capitalist countries of the core and subsequently in their relations with the periphery (Dear and Scott, 1981; Kirk, 1980; King, 1977)

Second, as the dynamics of global restructuring developed through the 1970s, the generation of 'urban planning knowledge for the periphery' began to assume a new role in the cities of the metropole. In the 1950s and 1960s, it had been an example of metropolitan dominance, as outlined on p. 49, with the cities in the colonial and immediately post-colonial periphery still manifesting the phenomenon of 'dependent urbanization'.

However, with the disappearance of industry from core cities (such as London or New York) and the maturing of the new international division of labour in the 1970s (see King, 1989a), the competitive advantage of such core states and cities in the world-economy has been maintained by a steadily increasing shift to the quaternary sector, to the knowledge-based, advanced producer services where superiority in information technology provides a significant instrument of control. Whilst most attention has been given to charting the rapid expansion of banking, insurance, international law, real estate, advertising, or management consultancy, and the monopoly of 'global control functions' in such cities, the growth of research and higher education functions in these cities has not been adequately charted (see King, 1990). Yet in the last ten or fifteen years, advanced education has assumed an increasing importance in the economies of core cities, and competition to increase the share of world markets in higher education (including, and especially from the 'developing world') has become a characteristic feature of core educational institutions since the 1970s. The question, therefore, is not what the ideological content of urban planning is today (whether in consultancy, research, graduate and professional training, publishing, or the policy orthodoxies of the World Bank, based in Washington) but rather, *where* its production and dissemination take place. In the new international division of labour of the 1980s, the 'dependent urbanism' is of the core on the periphery.

Chapter Four

VIEWING THE WORLD
AS ONE (1)
Urban history and the world system

INTRODUCTION

There would be little disagreement that one, if not the major change in British cities since the war has been in the ethnic composition of their populations. The Black and Asian population of Britain were estimated at some 2.1 million in mid-1980 by the Office of Population Censuses and Surveys, and formed 3.9 per cent of the total population, compared with under 1.4 million (2.5 per cent) in 1971, and some 200,000 in 1951. The concentration of this population into four or five major conurbations where, until economic crisis brought unemployment, there was a high demand for labour, means that anywhere between 1 in 10 to 1 in 20 inhabitants of many towns is of so-called 'New Commonwealth' and Pakistan origin; in certain London boroughs, depending on definition, the figure was 30 per cent in 1980 (Runnymede Trust and Radical Statistics Group 1980: 1) and in particular districts such as Shoreditch, Whitechapel and central Brixton, nearer 50 per cent by 1986 (Elliott, 1986: 15).

Extensive postwar immigration has, of course, not been confined to British cities. For Britain, France, and the Netherlands, the movement of labour has been from colonial or ex-colonial societies; for Germany, Switzerland, and France, it has also been from the Mediterranean. Between 1945 and 1975, a mere thirty years, some 15 million immigrants and their families have settled in Northern Europe, largely coming from the Mediterranean, a figure that might be compared to the 37 million arriving in the United States in the 100 or more years between 1820 and 1927 (Power, 1979: 1; see also Cohen, 1987).

This international movement of labour has perhaps been the most visible aspect of the workings of the new international division of labour. Hardly less visible has been the disappearance of jobs in the 'advanced' economies or their transfer to low-wage countries of the Third World, as well as changes in the physical fabric of the city, resulting from the restructuring of capital at both a national and international level. Here, the growing global influence of the trans-national corporation has become a major factor affecting national economies since the war. To quote from one account, 'of the hundred largest economic units in the world, only half are nation states: the rest are multi-national companies' (Makler, Martinelli, and Smelser, 1982: 13; Cavanaugh and Clairmonte, 1984).

In short, it is common knowledge that the economy of the country in general and of her cities in particular depend today on fluctuations in a capitalist economic system that operates on a global scale. The economic fortunes of Tyneside depend on decisions taken in Tokyo just as those of urban Scotland are affected by policies made in corporate headquarters in New York. As a contributor commented at the 1982 World Congress of Sociology in Mexico City, Germany's largest industrial city is São Paulo, Brazil. As a place of production, consumption, administration, or culture, the city is embedded in a global economy.

Yet if we look, in the history of urbanization in Britain, or especially in the history of individual cities, for an understanding of how this situation came about, with some exceptions, very little can be learnt. On the whole, relatively little attention is devoted to seeing that British urbanization developed as part of a larger international process. Leaving aside for the present the question of international capital flows, the rise and fall of world markets, the derivation of raw materials and energy sources on which the modern city was built, one might look only at the question of their changing ethnic composition.

Here, many of the sociological studies of ethnic minorities take as their starting point emigration into Britain from the 1950s, with greater or less attention given to the colonial background; Rex's work on race relations, discussed on p. 80, located in a framework that draws on global data and within a sociology of empire is a major exception (Rex, 1981; 1982). Despite Hobsbawm's (1969) suggestion of twenty years ago (in what in retrospect must seem to be an uncannily prescient phrase) that Britain was part of a global economy and could only be understood in such a context,[1] the framework for many, if not all, urban historical studies is still predominantly national even, or perhaps especially, in recent comparative work. Whilst this is a fairly sweeping generalization, its validity must be tested in the context of the argument being presented here. With regard to cross-national, comparative studies, it is in their essence that data is organized on a national basis in order to make comparisons, rather than in the form of international flows of capital, labour, or ideas that would illustrate processes of a different kind. As a fairly representative sample, mention might be made of a collection of essays on the origins and development up to 1974 of one of Britain's largest industrial cities, based mainly on engineering and textiles, that discussed the city's history principally in terms of local, regional, and national data. The 20,000 or more inhabitants of Asian and West Indian origin who were estimated to live there in the mid-1970s were not mentioned, nor was there any discussion of the origins of raw materials or of markets outside the UK.[2]

It could be argued that urban history in Britain had its genesis (in the early 1960s) in a different set of problems and concerns, some of them local and 'environmental', than those that beset the city today. Its paradigms were drawn from a different view both of the city and the world. Yet this approach to the historical understanding of British cities is in stark contrast to the burgeoning body of urban research on Africa, Asia, and Latin America in which modern urbanization in general, and the development of individual cities in particular, is understood in terms of their incorporation into a capitalist world-economy. If the reality of urbanism and urbanization in these societies, whether it is concerned with employment structure, capital investment, urban form, built environment (including 'squatter settlements'), political process, ethnic composition, or any other topic is to be understood as a result of 'the penetration of the tropical world by finance capital' (Lloyd, 1979: 15), their fortunes dependent on decisions made in the core countries of European and North American capitalism, it would seen equally logical to see the development of these core cities, and particularly the industrial cities of Britain, as being related to other parts of the world (and not only the Third World). That British cities have not, until quite recently, been understood in this way would seem to result from three phenomena: the problem of disciplinary specialization; related to this, the different and often uncomplementary sets of paradigms with which different disciplines often operate; and the concentration of much urban history on geographically defined areas rather than on economic, social, political, or cultural processes operating on a global scale.

What this chapter attempts, therefore, is to review some of the theoretical developments regarding cities and the world-economy that have taken place in the last few years with a view to assessing their relevance for urban historical research in Britain.

URBAN STUDIES AND DEVELOPMENT THEORY

Two different, though frequently overlapping areas of discourse have, from the late 1960s, undergone massive changes: the first in urban sociology, subsequently and alternatively identified as urban political economy or neo-Marxist structural approaches, and the second, development theory, as it moved through theories of social change, modernization, dependency, world systems, and the internationalization of capital (see Chapter 1).

The paradigm shifts in urban sociology were sufficiently well charted in the 1970s and early 1980s for any lengthy restatement of their intention to be unnecessary (e.g. Aiken and Castells, 1977; Castells, 1977; Harloe, 1977; Harloe and Lebas, 1981; Mingione, 1981; Pickvance, 1976; Tabb and Sawers, 1978; Walton, 1979; 1984; Zukin, 1980) and have been continued in *Society and Space* (see especially, 1987: 4) and elsewhere. Much of this work developed in association with the Research Committee on Urban and Regional Development of the International Sociology Association established in 1970, of which the 'house journal', the *International Journal of Urban and Regional Research* began publication in 1977. In that year, Aiken and Castells described some of the characteristics of the new trends in urban research as (1) examining the larger social, economic, and political contexts of cities; (2) using an historical perspective to study urban problems and phenomena, meaning a strong emphasis on the process of social change over time in urban systems; and (3) exploring the critical role of the economic system in shaping the nature of urban systems.

> More specifically, a fundamental postulate (underlying) these new approaches is that a given element of an urban system cannot be properly isolated as a separate object of study, meaning it cannot appropriately be removed from the economic, political, social and historical context of which it is inextricably a part.
>
> (Aiken and Castells, 1977: 7)

71

Instead of reifying the town or city as a self-contained object, treating it as some independent entity or, in an explanatory context, as a variable, it was rather to be seen as an arena where larger economic, social, and political processes were worked out.

From a given Marxist perspective, urbanism was viewed as the particular geographic form and spatial patterning of relationships taken by a particular mode of production and the process of capital accumulation. According to this viewpoint (and drawing on Hill, 1977), this required: fixed investment of part of the surplus product in new means of production; the production and distribution of articles of consumption to sustain and reproduce the labour force; stimulation of an effective demand for the surplus product produced; and additional capital formation through ever-increasing product innovation, market innovation, and economic expansion: 'the capitalist city is a production site, a locale for the reproduction of the labour force, a market for the circulation of commodities and realisation of profit, and a control centre for these complex relationships' (Hill, 1977: 41; see also Hill, 1984).

A useful overview of the state of the art about this time was provided by Walton (1976) who brought out the problematic concerning the nature of 'the urban', the need to understand urban forms by reference to changing modes of production, with the central feature of the contemporary (capitalist) city being the growing concentration of the means of collective consumption and the organization of production and consumption. As Smith summed up much of this work in 1980, 'urbanisation is often confused with capitalism. The effects of capitalist economic development often are mistaken for effects of urbanisation' (Smith, 1980: 235). It was not that this conclusion was especially new; rather that a decade of research had helped to forge the theoretical tools with which the processes could be understood. The absence of commitment to an orthodox Marxist position was not an impediment to appreciating the significance of these developments, nor did others forget the history of pre-capitalist urbanization (see especially, 'Introduction' in Abu-Lughod and Hay, 1977; Friedmann and Wulff, 1976; Walton, 1979).

Yet despite the international context in which the 'new urban studies' were undertaken, much of the work was, till quite recently, as Walton pointed out in 1976 (p. 307), *intra*-national in focus; nor did these or more conventional urban studies add up to a more coherent theory of urbanization (Friedmann and Wolff, 1982). Little had been done to extend the analysis 'to cross national urban hierarchies or world urban systems despite the fact that the fundamental process of concern is clearly, in Amin's (1974) words, one of 'accumulation on a world scale' (Walton, 1976). (The reference is to S. Amin (1974) *Accumulation on a World Scale.*)

If the work published in the *IJURR* is taken as representative, the intentions of the 'new urban studies' had not been fulfilled at the beginning of the 1980s; most studies concentrated on the core capitalist countries of Europe and North America, with little work on Asia and Africa; there had been few empirically grounded historical studies; for researchers interested in the built environment (as are many historians), there had been few studies (with the major exception of Harvey's work (Harvey, 1973; 1985) and, because of its subordination by economistic perspectives, cultural variation, whether as a dependent or independent explanatory variable, had tended to be a non-starter.[3] From the mid-1980s, for a variety of reasons, this situation changed (King 1990).

It is with the development of the world political-economy approach to urbanization, welding together the perspectives of the two fields outlined previously, including notions of the world-system (Wallerstein, 1974; 1979; 1984) that more promising developments have taken place. Where previously:

> development research guided by the modernisation perspective tended to concentrate exclusively on problems internal to Third World countries . . . more recent theoretical literature redefined the concept of development as a process embedded in the structure of the world economy and having consequences for both advanced and backward societies. Hence, the study of development should not be limited to underdeveloped countries but it should include, as a primary concern, structures and process in the international system.
>
> (Portes and Walton, 1981: 3)

From this research, two early concepts or ideas have stood up well over the years, first, Castell's notion of dependent urbanization, and second, in the context of capital accumulation on a global scale, ideas stimulated by Harvey's oft-quoted paragraph:

> the geographical pattern in the circulation of surplus can be

conceived only as a moment in a process. In terms of that moment, particular cities attain positions with respect to the circulation of surplus which, at the next moment are changed. Urbanism, as a general phenomenon, should not be viewed as the history of particular cities, but as the history of the system of cities within, between, and around which the surplus circulates . . . the history of particular cities is best understood in terms of the circulation of surplus value at a moment of history within a system of cities.

(Harvey, 1973: 250)

Whilst originally, Castell's notion of dependent urbanization implied that urbanization took place in the colonial or neocolonial society but the industrialization, which was historically associated with urbanization in modern societies, occurred in the metropolitan (Castells, 1977: 47–9), in more recent years, it has become evident that it is metropolitan urbanization that is equally dependent.

Implicitly or explicitly, this latter concept has been stretched and extended to cover a large variety of circumstances in which urban processes at the periphery of the development of the world-economy are understood only in relation to processes at the core. Thus, the fortunes of colonial cities such as Rio de Janeiro, Cape Town, Algiers, or Calcutta have been related to the particular times and circumstances in which they were incorporated into the world-economy, and their function as well as organization and spatial structure understood in this context (Ross and Telkamp, 1985). Likewise, the spatial organization of society, or the sectoral distribution of the workforce between agriculture, industry, and services in the nineteenth century in Latin America or Australia, is seen as structurally related to the prevailing economic systems and social formations in the metropolis (Salinas, 1983; Roberts, 1978; Browning and Roberts, 1980; Mullins, 1981). Other research agendas suggested examining cities according to their role in the accumulation process, with historical research focused on the rise and fall of cities as a result of changes in the nature of the circulation of surplus, or with regard to the distinctive functions they serve in the exchange process (Walton, 1976).

In the context of colonialism, however, two or three observations can be made. First, individual empires have tended to be seen as self-contained entities, rather than as parts of a larger process; second, the definition of what is colonial has tended to focus on what might be called 'successful' examples of colonialism, i.e. those of what are now termed Third World countries, recently independent, and not to have treated within the same framework other one-time colonial societies (such as Australasia or the North American colonies) (see Chapter 3). However, urban historical work on Australia has adopted a world political-economy framework in which the distinctive urban experience of that country has been understood in relation to a developing international market system (Mullins, 1981; King, 1984).

Wallerstein has described the world-economy as 'a single division of labour within which are located multiple cultures'. Essential elements other than these are: production for profit in a world market, capital accumulation for expanded production as the key way of maximizing profit in the long run, the emergence of three zones of economic activity (core, semi-periphery, and periphery), and the development over time of two principal world class formations (the bourgeoisie and the proletariat), whose concrete manifestations are complemented by a host of ethnonational groups. This historically unique combination of elements first crystallized in Europe in the sixteenth century and the boundaries slowly expanded to include the entire world (Wallerstein, 1979: 159).

Whilst the use of the world-system perspective increased in the 1970s, there was still, in 1981, according to Portes and Walton, 'a manifest disjunction between general theory, where the world system perspective has become dominant, and the myriad lower level focused studies — national and thematic — based on the earlier modernisation model' (Portes and Walton, 1981: 13).

CITIES AND THE WORLD-SYSTEM

However, two valuable 'agendas for research' emerged about this time (Walton, 1979; Friedmann and Wolff, 1982). Walton's (1979) proposals for research directions for a comparative political economy would all seem to be relevant for urban history. In order to appreciate what is unique about urbanization under advanced capitalism, it was essential to know more about the experience of socialist and peripheral capitalist states; in the context of urbanization under peripheral capitalism, more attention needed to be given to the historical, political, and economic circumstances accompanying the incorporation of such peripheral societies into a world-economic system and how precolonial and colonial urban centres were transformed in the process; urban social movements, nationalist revolts, and urban reforms were also important

75

to investigate. Most important, however, were his suggestions for research on cities and the world-economic system.

Cities of the advanced capitalist world depend on Third-World markets for both raw materials and often, migrant labour. The economic survival of certain capitalist cities may even depend on this exchange, just as the less potent urban centres like textile manufacturing towns might die out with the export of industries to Third World cities, where costs of production were lower and profits higher. Likewise, poverty and unemployment in Third-World cities were closely linked to production for export to metropolitan centres. The extent of change in the perception of urban change in core countries in the last decade is evidenced by other comments of Walton's in 1979: 'This latter fact has become a central concern in research on the underdevelopment of third world cities and societies. The reverse is not true'. For Walton at the end of the 1970s, the most serious failing of the new urban political economy studies was its inattention to the international networks within which cities were located and shaped (Walton, 1979: 12–14).

It is these objectives that have been appreciably advanced in the 1980s, not least in terms of historical research. In addition to the studies of Browning and Roberts (1980) already mentioned, Slater (1980; 1986) has been concerned to develop a framework within which upturns and downturns in the global economy influenced the internal development of regions and classes in particular areas. For studies that centre more on British concerns, Rex (1981: 182), while drawing selectively on Wallerstein, gave more attention to the distinctive political and cultural context of specific empires. However, a number of frequently cited studies clearly established the way cities are linked to world-system processes — Cohen (1981) on the connection between the new international division of labour, multinational corporations, and urban hierarchy, Portes and Walton (1981) on the emerging 'backwardness' and unemployment in advanced countries, especially the USA, resulting from the flight of capital and the increasing internationalization of capital and labour processes resulting from large-scale migration, and Friedmann and Wolff (1982) on world-city formation. In looking at world cities in a world-economy defined by 'a linked set of markets and production units organised and controlled by transnational capital' Friedmann and Wolff (1982) set out what they termed 'an agenda of ignorance', which raises many questions of a historical nature, for example: on the origins and development of transnational corporations as 'the chief instruments for the globalisation of the economy'; on the effects of technological change; on the effects of emerging global markets and production units on employment sectors and particular profiles of urban development; or on the predecessors of the new 'transnational elites' who manage 'the transition to the world economic order'.

Friedmann and Wolff (1982) argue specifically for a historical approach that focuses on the transition of particular cities to world-city status (as 'control centres') in the context of an expanding world capitalist economy. Whilst these suggestions emphasize the economic dimensions of change, they provide a framework for examining social and cultural phenomena (not least the ethnic and racial composition of cities), the local and national resistances to the processes they chart and growing interest in the construction of subjectivity.

SOME SUGGESTIONS FOR URBAN HISTORICAL RESEARCH

The major contribution of world-system studies has been their perspective, 'the belief that something is going on above and beyond individual societies' (Bergesen, 1980: xiii; Cohen, 1981). Though utilizing certain common concepts (core-periphery, international division of labour, or the cyclical boom-bust of the capitalist world-economy) individual studies obviously show considerable diversity in both approach and method. Moreover, it is also clear that these approaches have various intellectual antecedents from Hobson's (1948) work on imperialism to the research of Brinley Thomas (1972) on the Atlantic economy, or Hobsbawm's *Age of Capital* (1975) which, among other things, pointed out that the term 'world economy' (in German) was already in use in 1870. Yet on the whole, as Walton suggested, literature on world-level processes of change has tended to develop in relative isolation, accumulating hypotheses at a highly abstract level but deprived of contact with, and potential insights from, close-range empirical investigation (Portes and Walton, 1981: 19). From the mid-1980s, however, such empirical studies have begun to appear (e.g. Henderson and Castells, 1987; King, 1984; Smith, 1984; Taylor and Thrift, 1982; Timberlake, 1985; Smith and Feagin, 1987; Walton, 1985).

It might be argued that the strengths of urban history in the UK have been precisely in this latter area though, as suggested on p. 69, the larger international economic, political, and cultural reality in which the city and society are embedded have been too frequently taken for granted. Where urban processes have been increasingly seen to be part of international processes, this view might be extended to a more genuinely global scale. The obvious problem is that whole new sets of data of a different nature will need to be collected.

The question is whether the real development of London or Manchester can be understood without reference to India, Africa, and Latin America any more than can the development of Kingston (Jamaica) or Bombay be understood without reference to the former. Nevertheless, the real divisions of scholarship, as well as the ideological underpinnings that help to keep them alive, ensure that histories of 'First'-, 'Second'-, and 'Third'-World cities are still kept tidily apart.

If, for example, explanations about the built environment are considered within a world-system perspective, we are confronted with the contradiction that, in the realm of architecture, the high-energy, high-technology, imported concrete monoliths imposed on colonial and post-colonial cities as part of the installation of modern capitalism are said to be in the 'Internationalist' style. It might, in fact, be worth making a digression to look at the terms 'international' and 'internationalism', to see the historic context in which they arose.

According to the *Oxford English Dictionary* (1971), the term was first coined in 1780, in the first major phase of modern colonial expansion: the term 'internationalise' was to develop a century later. The examples given by the *OED* are instructive: the meaning given 'to render international in character or use or in politics, to bring (a country, territory, etc.) under combined government, or protection of two or more different "nations" ', for example, from the *Contemporary Review* (1883), 'an earnest appeal to the Government at Berlin to unite with England to internationalise the Congo'; from the *Spectator* (1885), 'The Suez Canal must be internationalised and confided to the Khedive'; hence, 'internationalisation' means 'the action of internationalising', for example E. Dicey (1882): 'The internationalisation, if I may use the word, of Egypt' or in 1884, where *The Times* refers to 'the internationalisation of the Congo, the Niger and other fields of commerce'.

One problem needing further clarification is the relationship between the overarching notion of the world-system and the specific role of historic empires within this. Though clearly part of a larger global political economy, each has had very different social and cultural dimensions that have mediated the urban experience both in metropolitan as well as colonial societies: that cultural influences in Britain are from India, or immigrant labour from Pakistan or the Caribbean, rather than Indonesia or North Africa is not unimportant.

Second, where considerable attention has been given to the role of transnational corporations in the world system, further research is needed on their historical antecedents, not only companies, from the East India Company (founded, 1600) to Unilever, but also banking, insurance, and, significant from the viewpoint of the transfer of technology and urban planning, to the role of military engineering and science (King, 1984).

Third, 'the circulation of ideas in the world system has received much less attention than the circulation of either commodities or capital' (Portes and Walton, 1981: 15). Though here, ideas about inequality are being referred to, an equally important field is the particular forms of built environment that have accompanied the expansion of capitalism round the world and the ideologies that both rationalize and support it. Here, Robertson's work on the process of globalization (Robertson and Lechner, 1985; Robertson, 1987) provides an alternative and complementary perspective to the more economistic world political-economy approaches.

Transnational capitalism is not only a particular way of organizing labour and capital; it is also 'a set of ideas about the world and a global community of people who subscribe to them'. This transnational community is made up of people who, though belonging to different nations, nevertheless have similar values, beliefs, and ideas 'as well as similar patterns of behaviour as regards . . . housing, dress, con-sumption patterns, and cultural orientations in general'. It is also represented in cities by a distinctive spatial organization with local elite communities concentrating in suburban residential areas and repro-ducing (in so-called developing countries), the urban structure, housing style, and architectural lay-out of the core countries (Sunkel and Fuenzalida, 1979: 68–75).

As part of the penetration of Third-World societies by the market economy, transnational enterprises have transformed societies and cultures whether through book publishing, educational systems, or consumption patterns in general; their role has been to create, in the words of Kumar, 'culturally relevant markets for Western goods'. Third-World elites have learnt 'to define development in terms of capitalist criteria rather than with reference to those standards which are rooted in their own cultural system' (Kumar, 1980: 166, 248).

The historical restructuring of the physical and spatial environment or, in plainer language, the transformation of architecture, dwelling forms, urban and environmental form, is crucial to the process that Kumar (1980) describes and ranges from changing patterns of land-ownership to the use of new (and often imported) materials.

The physical, spatial, and technological transformation of the domestic as well as the larger urban environment is, after the transfer of land-ownership rights, the most fundamental prerequisite for creating the city as a place of consumption. The export of capital in the nineteenth and early twentieth centuries was, in many cases, accompanied by the export of the 'intellectual technology' of British expertise, whether with regard to railway systems, urban planning, or municipal legislation (Christopher, 1988) (just as today, oil-rich states of the Middle East rely on Western intellectual technology, whether with regard to urban planning or the training of soccer teams, to help invest and circulate surplus capital).

Fourth, it is clear that certain segments in the built environment in the UK can easily be seen not just as the product of a global economy but of a real international division of labour. If, 'in a very real sense, the slave plantations of the Caribbean . . . were part of the social structure of modern Britain' (Rex, 1981: 3), the vast country houses and estates such as Harewood, or Fonthill, built on plantation profits (Williams, 1944: 87–97), or on capital accumulated from India (Holzman, 1926; King, 1984: 66) were part of an urban system created by the capitalist world-economy. Other areas, however, were equally part of that system, from the rubber, sugar, and metal-processing regions to the Lancashire cotton towns. What is the *total* explanation for the fact that one of these, i.e. Blackburn, had, in the 1980s, one of the highest proportions locally of Asian residents?

Apart from these more obvious spheres, however, two aspects of British urbanization seem particularly appropriate for examining in this larger context: the rise and subsequent decline of the coastal leisure resort, a classic case of the investment and circulation of surplus capital, and late-nineteenth- and early-twentieth-century suburbanization. In the first case, there is a need to examine not simply the local and national sources of investment and patronage but also, the international ramifications of accumulation on which the whole depended (King, 1980; 1984: 65–67). The present multimillion dollar industry of time-sharing and second-home development where, as part of a global tourism industry, multinational contractors transfer their building activities from a depressed domestic market to the Canary Islands or the Caribbean (O'Neill, 1982) has an earlier history in railway building and hotel development in the south of France and Switzerland.

A second area is in the realm of suburbanization. The development of an effective international division of labour from the third quarter of the nineteenth century, with some economies (such as Britain) becoming increasingly industrial and others agricultural, was to bring a decline in land prices and the wholesale disposal of land before and after the First World War as the country became increasingly dependent on overseas food. Lower land prices combined with transport innovations, the shift (especially in the South-East) to services, and other factors meant not only immense suburbanization in the post–First World War period but, with the huge emigration of labour before the First World War (almost three-quarters of a million people in the first decade of the century left for Canada, Australia, and the informal empire of Latin America), the transfer of particularly English suburban ideas of 'detached' domesticity abroad (Samson, 1910; King, 1984: 91–4). As Christopher (1988) states, migration overseas was principally urban to urban. It is not only in terms of the capital–land relationship, however, that suburban development needs to be understood in a world-economy context but also, in terms of a variety of crucial products, not least oil, rubber, timber, or asbestos, which a particular colonial mode of production made available for the development of suburbanization in the interwar period.

In the early stages of this chapter, reference was made to the urban history of a particular city; it will conclude by referring to three phenomena in that same city, each of them concerned first with understanding, but also, questions of policy and action. In the early 1980s, the main neighbourhood of housing previously owned and maintained, in a culturally distinctive style, by residents of Asian and Caribbean origin, was selectively demolished and its inhabitants decanted elsewhere. Where conservation policies place high value on traditional bourgeois areas, the real cultural innovations, often making the most significant contribution to the variety of postwar cities, and genuinely expressing the pluralism which is their most conspicuous social fact, were subject to municipal destruction and neglect. Rather than be declared conservation areas, housing in such areas was demolished, the vacant lots sold for the display of adverts for consumer commodities (such hoardings, of course, being strictly banned from bourgeois conservation areas). It is evidence of the changed perceptions of the 1980s, that these policies were subsequently changed (though not in relation to conservation).

Further into the centre of the city, an 'industrial museum' was established that traces (because the larger part of it has disappeared) the development of the woollen-textile and ready-made-clothing industry in what, at one time, was probably one of the largest centres of this kind in the world. In the museum, there is no reference to the reasons for the industry's decline, nor to the countries (perhaps Taiwan, Portugal, Korea, or Hong Kong) where the cheaply produced clothes worn by the visitors to the museum are probably made. Though the immigrant Jewish population figures prominently in the history, no mention is made of the most recent immigrants.

Finally, in the window of the modest estate agents five minutes from where this piece was written, along with photographs of 'back-to-backs' and 'semis', were advertisements for villas on the Costa Brava as well as time-share apartments and ranch houses in Florida. 'The world economy is everywhere' (Friedmann and Wolff, 1982).

All cities today are 'world cities', yet they have not just assumed that role overnight. The agenda for urban history that perceives them in this way is clearly vast. Yet such a perspective would enable urban problems, economic, social, and physical, to be seen in a much more realistic light. Not least from an educational perspective, it could not only help create new attitudes and knowledge but provide a more informed insight into problems of race, unemployment, and physical renewal. The cosy viewpoint of looking at our cities from within must be replaced by the more uncomfortable view of seeing them from outside.

Chapter Five

VIEWING THE WORLD AS ONE (2)

*Culture and the political economy
of urban form*

INTRODUCTION

The question of deciding on 'appropriate standards' for housing,
building, and planning — of 'norms and forms' to use Rabinow's well-
chosen expression — in the 'developing', ex-colonial countries of the
periphery is a topic of fundamental importance to planners, architects,
and administrators, and it has been frequently discussed (e.g. Hardoy
and Satterthwaite, 1981; Mabogunje, Hardoy, and Misra, 1978;
Payne, 1984; United Nations, 1971).

In these and similar discussions, reference is often made to the
inappropriateness of imported or 'foreign' standards for Third-World
housing and planning. Official standards are said, for example, to rely
on 'Western technology and social philosophy' (Mabogunje, Hardoy,
and Misra, 1978: 8), to promote 'Western-type building materials and
techniques' (Hardoy and Satterthwaite, 1981: 256) and that building
codes emanating from a bygone colonial era prevail or are promoted
in developing countries (United Nations, 1971: 36). In contemporary
discussion, whether lay or professional, reference is made — for
example, with regard to building development in Kuwait — to
'Western-style re-development' (Maclean, 1982: viii) and in a current
textbook on *Cities of the World* (1983) the authors refer to 'the dilemma
of Westernization vs modernization'. This they spell out in more detail:

> change (in less developed countries) is likely to take the form of
> Westernization, since it is the West, the more developed countries,
> that developed the lead in creating the modern industrial city and
> the life-style that goes with it. To be sure, there are ample signs
> of this Westernization (some might call it 'homogenization' or

'internationalization') of the world's major cities, in the form of skyscrapers and modern architecture, the automobile society, Western high technology, advertising, *etc.*

(Brunn and Williams, 1983: 36)

These shorthand labels, 'Western', 'modern', 'international', tend, in everyday language, to be used in a blanket fashion to cover a wide range of phenomena — not simply 'architectural style' in its most limited sense of appearance or facade, but also building and urban form, materials, environmental standards, or building types, to say nothing at this point about the institutions (economic, social, and political) that the building or built form contains. What I would like to attempt in this chapter is to look at these and other terms, the categories and labels with which built form is often thought about, and to ask some related questions. In particular, it will address three issues.

1. What are the appropriate labels to describe built form and the larger built environment, particularly in the context referred to previously? How accurate or appropriate are terms such as 'Western', 'modern', 'European', 'international', or 'post-modern'?
2. If we accept for the present one or two of these terms, *i.e.* 'Western' or 'modern', to describe the metropolitan or core society initially involved in the historical, and often colonial situation, in transplanting Western standards into particular Third-World settings, what explanation can be offered for the norms and standards of housing, planning, and urban development prevailing in that core society?
3. When such standards and norms were or are introduced into the Third-World society (*i.e.* the process of 'Westernization', 'modernization', or alternatively, 'bourgeoisification', what are the processes at work?

One aim of *Colonial Urban Development* (King, 1976) was to examine the impact of 'Western industrial colonialism on urbanization and urban development in non-Western societies', particularly India, and to develop a theory and method to do this. Very briefly, the argument suggested that the built environment resulted from a particular form of political economy, namely, colonialism; how that form was expressed, however, depended largely on the society and culture of the countries concerned. Whilst the particular institutions of colonialism were discussed, a considerable part of the analysis was devoted to analysing culture in terms of values, activities, relationships, and institutions and how these were subsequently related to built form.

Explanations, however, theoretical or otherwise, emerge from consciousness, and this is perhaps the first point to emphasize. In this case, the social and cultural consciousness had been a product of some five years in India, subsequently fostered by prevailing theoretical work in the early 1970s (particularly, the pioneering work on cultural variables by Amos Rapoport, 1969) as well as my own research findings. An anecdotal digression might be appropriate at this point to discuss these.

Part of the research for that study was undertaken in the rich and fascinating archives of the India Office Library, particularly, the Western Drawings Section with over 10,000 paintings and drawings and innumerable photographs. It should also be stated that prior to going to India, not only had I no knowledge of this collection but also, was largely unaware of the researches of cultural geography or the large amount of research in anthropology and other fields on cultural variables in perception and similar issues. In addition, having grown up as part of an 'art-conscious' family, I had positive — if yet modest — views about my abilities as a photographer. Hence, during those years in India, many photographic records were made varying from 'nice views' and 'memorable occasions' to 'more objective' records of buildings and environments of professional interest. To discover, on returning to Britain, a library of thousands of drawings, paintings, and photographs of India made by earlier British travellers and residents was, therefore, both a pleasure and a shock. Not only had British predecessors, over a period of more than two centuries, been making records, according to certain criteria, of what they thought was important, but more particularly, there were very clear continuities in what they perceived and how they had both recorded and presented it. This was also reflected in how and what they had built (Figures 5.1(a) and 5.1(b)). More sobering, however, for one's ego, was the discovery that more than a few of my own 'creative photographs' were virtually identical in subject matter and composition to drawings or paintings made by unknown 'cultural and social forbears' 80 or 100 years previously (Figures 5.2–5.5).

If not the most important, this is the most easily illustrated example

Figure 5.1(a) Painting of house at Madras, early nineteenth century (courtesy, India Office Library)

Figure 5.1(b) Photograph of Commander-in-Chief's Residence, New Delhi, 1942 (courtesy, India Office Library)

Figure 5.2 'Looking over Jumna, Delhi, 1903' (courtesy, India Office Library)

Figure 5.3 Cochin, Kerala, 1970 (author's photograph)

Figure 5.4 Central India, *c.* 1860. Painting (courtesy, India Office Library)

Figure 5.5 New Delhi, 1970 (author's photograph)

of what, in effect, were continuities not only in social and cultural practices (in this case, perceptual criteria and normative activities) but also in much more fundamental and structural factors governing social organization and culture and their relation to the physical and spatial environment. Whatever understanding of the phenomenon of 'colonial urban development' in India came from an examination of the historical and contemporary data, many insights came from the structural continuities, economic, political, social, or cultural, derived from participant observation. In this context, therefore, the emphasis on 'social and cultural' as opposed to more 'political economic' interpretations (though the distinction between these two levels of analysis is clearly nowhere so clear-cut as this sentence implies) is perhaps understandable (see also Walton, 1984).

The question arises, however, as to how social organization and culture, understood in terms of values, beliefs, attitudes, ways of doing things, are themselves formed and changed; in particular, how are social relations modified by the cash nexus, how are human needs commodified and social institutions changed by the development of market relations? Here, attention might be refocused on the emergence of industrial capitalism in the West, and particularly, Britain. The following generalized comments are offered more to illustrate a type of argument than to prove it.

Although there are many studies of the development of the industrial city and, in a larger sense, the restructuring of the physical and spatial environment following the emergence of a capitalist mode of production, the way in which the specific forms of this were related to the particular context and processes of that mode of production is an area on which there is plenty of scope for further research. Nor, presumably, will more insights be possible until more is known about the real or theoretical possibilities of the development of industrialization under non-capitalist conditions that develop, *ab initio*, with the state ownership of land, property, and the means of production and that would, presumably, lead to a totally different kind of built environment. In other words, in the historical context, what aspects of the built environment were structured by 'the capitalist mode of production' as it has subsequently been introduced into other societies, what aspects resulted from the distinctive social/geographical/cultural conditions of its introduction and third, what aspects resulted from the particular historical or situational context governing the available technology, energy system, or the like?

In a very general sense, a mode of production, or system of socioeconomic organization is expressed both spatially as well as in the form of the built environment. There is both a social and spatial division of labour that historically, in the case of Britain (Massey, 1984), required not only a concentration of labour in urban areas but also an increasing specialization in the urban hierarchy, with some towns and regions concentrating, for example, on textiles or engineering, others on mining, and others, such as London, on banking and insurance (see also Lash and Urry, 1987). This spatial division of labour was equally reflected in the built environment, whether in terms of the general distribution between urban and rural areas or in its specific concentration, both at the place of production (the factory) or in the accommodation constructed close by (Chapman, 1971; Tarn, 1971; Caffyn, 1986).

Apart from this modified economic, social, and spatial order, related as it was to historically specific forms of energy production, are new forms of economic and social organization: two of these have been mentioned — the factory, and accommodation for labour, which led to the emergence of new built forms. Similarly, other new forms of economic and social organization, whether related to technical or energy criteria, are reflected in new building and settlement types — the engine shed, the mill, the mining village, the bank. The increased sophistication of economic organization leads to new organizational forms: insurance transactions that once took place in a coffee house later occupy a complex specialized building; the introduction of the concept of limited liability modifies the form and scale of a stock exchange; specialization in industrial and economic processes leads to an increasing variety of industrial building types. With these changes come modified forms of social organization and social control that — irrespective of Marxist or Foucauldian perspectives — have resulted in the production of new and specialized institutions reflecting the redistribution of labour between rural and urban locations: the workhouse, the prison, the asylum, the hospital, as well as an increasingly differentiated provision for education and government, all of which are associated with the development of the State (Evans, 1983; Foucault, 1973; 1979; also articles by Forty (1980), Scull (1982), and Tomlinson (1980) on the hospital, the asylum, prison, and 'Introduction', in King, 1980).

A basic characteristic of capitalist industrialization, however, and of the rural–urban division of labour that distinguishes it from

subsistence or peasant economies, is the size of the surplus production and its selective appropriation. According to Harvey, the urban
process 'implies the creation of a material physical infrastructure for
production, circulation, exchange and consumption', with the built
environment acting as a principal means of accumulation and opportunity for keeping the surplus in circulation (Harvey, 1985). In terms
of historical analysis, insufficient attention has been given, in this
context, to the function performed by what today is referred to as the
leisure industry but in conventional historical discussion is usually
referred to as the rise of the resort. What Hobsbawm, for example,
shows is the correlation between the development of industrial production, the investment of finance capital, and the recycling of surplus
during the third quarter of the nineteenth century with the development of the leisure resort and 'conspicuous consumption' (Hobsbawm,
1975). In what was the first major construction boom, there was not
only a rapid expansion of leisure building in terms of resorts and spas
but also, the invention of new building forms and leisure environments,
whether the seafront promenade, winter gardens, or the specialized,
mass-vacation house (Hobsbawm, 1975; King, 1984). There was also
in this period, a spectacular increase in the production of architects
(the number more than doubled, from less than 3,000 to almost 7,000
in the thirty years after 1851 (see Kaye, 1960: 173)). This would seem
to provide the first historical precedent to contemporary investment
in the leisure industry with the proliferation of leisure centres, tourist
resorts, time-share apartments, or theme parks.

So far, however, these sketchy historical comments have focused
on developments in one society. What is, however, apparent is that,
at least from the eighteenth century, if not earlier, this new mode of
production was operating on a global scale. However, just at what
point in the nineteenth century one can accurately and comprehen-
sively speak of 'global relations of production' is arguable; nevertheless,
through the instrument of colonialism, the penetration of subsistence,
peasant, or semi-incorporated economies by the system of industrial
capitalism in which peripheral economies were, to a greater or lesser
extent, linked to those at the core largely occurred at this time.
Mabogunje, for example, writes of a 'more appropriate spatial order'
imposed on West Africa by the penetration of capitalist institutions:
a new urban system of ports, railways, and administrative towns that
effectively linked up the urban system of the colony with that of the
metropolis, as well as urban systems of one colony with those of

others (Mabogunje, 1980). Others have demonstrated similar processes occurring in North Africa, Latin America, India, or elsewhere (Abu-Lughod, 1976; Roberts, 1978; Bayly, 1983; Pang, 1983). Again, at exactly which point in time one can speak accurately, like Wallerstein, of the capitalist world-economy as 'a single division of labour within which are located multiple cultures' (Wallerstein, 1979: 34) needs to be traced in terms of built environments and urban forms, not least, the point at which the 'multiple cultures' begin to form one (see Chapter 1).

If, therefore, there has been a move towards a system of global relations of production, and if this is expressed both spatially (in the degree of urbanization) as well as in other aspects of the built environment, this provides a larger framework in which to analyse built form, both in the metropolis and in the Third World.

Ideas about the social and spatial division of labour at the regional and national level are combined with the notion of a division of labour at an international level that is also both spatial and cultural. In terms of understanding built form in both core and peripheral societies, including questions of standards, these concepts might provide better scope. Thus, historically, 'Western' or 'modern' parts of the built environment in the colonies, or today, in the Third World, are those that are associated with the managerial or organizational functions of the developing capitalist world-economy and are used by colonial or *comprador indigenes*, or subsequently, by indigenous elites linked to the world-economy. Indigenous labour — and those parts of the indigenous economy linked with pre-colonial modes of production — are those associated with traditional or 'cultural' parts of the built environment, manifesting 'cultural standards'. (The distinction between 'official' and 'cultural' standards is made by Mabogunje, Hardoy, and Misra, 1978.)

A simple, but useful example is that of the European or Western bungalow of the plantation manager and the cultural standards or traditional accommodation of plantation labour. Or, to take another example from the metropolitan or core society, the distinctive built form of the densely packed terraced or row housing of the Lancashire industrial cotton town (sic) must be viewed within a global political economy that has a physical and spatial dimension and includes the cotton exchange in Manchester, the transport links to the cotton supply (whether in the southern states of the USA, or in India, or Egypt) as well as the cotton plantation or fields and the accommodation of

95

those connected with it. The main point to be made here is that all aspects of built form — whether seen as spatial, physical, or architectural — must be viewed within an emergent, interdependent, and interacting capitalist world-economy, yet one that permits, in different parts, different cultural expression.

As indicated on p. 72, the historical role of the city — though not, of course, the only one — has been to act as a place of production and consumption. Over the last few years, the direction of critical theory in planning has been to see urban planning in capitalist societies as an activity concerned with rationalizing, or optimizing market processes, eliminating obstacles rather than, with certain qualifications, restructuring society. Where planning was 'exported' to the Third World in colonial or post-colonial situations, the situation was similar, though with the additional dimension that, prior to the national reorientation of urban and regional planning, the function was to rationalize urban forms and activities that were already linking the colonial society to the metropolis (see Chapter 3, this volume). These generalizations can be better understood in relation to three propositions concerning different dimensions of the process of urbanization.

1. Urbanization is said to be the production of capitalist development and expansion: the number and size of urban centres in developing countries relate to patterns of investment and consumption (Roberts, 1978: 11, 81).

2. In modern, industrial market economies, suburbanization involving high rates of home and car-ownership is seen as essential for expanding consumption and maintaining GNP (Harvey, 1973: 1985).

3. There is a particular form of suburbanization characteristic of modern capitalist economies (including urban structure, housing style, and architectural lay-out) that places high value on separate, single-family homes. The logical extension of these developments is towards an increasing proliferation of smaller households, each one of which is accommodated in its own separate, owned property.

These propositions bring us back to one of the questions stated at the outset: what is the relation between 'culture' and 'political economy'? In particular, how are values, expectations, and ways of doing things, changed by the historical experience of the operations of a market economy? In Britain, historical experience would suggest that increasing GNP and the operation of the market economy have led, during this century, not only to increased owner occupation but also, to the multiplication of separate households, a growing proportion of which have been contained in separate, single-family (detached or semi-detached) dwellings (King, 1984). Is the explanation of a particular housing preference (and hence, a particular kind of built environment) therefore, to be couched primarily in 'cultural' (i.e. British?) or 'mode of production/political economy' terms (i.e. free market/capitalist?) terms? Certain social and cultural values such as privacy seem to be especially related to the values of property ownership (Brittan, 1977).

A question might be asked about the large amount of research generated in recent years in the field of architectural or environmental psychology on the subject of 'privacy', and particularly in the market economies of the USA and Europe, and to what extent this is a product of the market economy and a concern with consumption. This is particularly relevant when considered in relation to building services: Kira, for example, finds it worth noting that 'even as recently as 1970, some hotelmen in Europe were proposing to build hotels without a private bathroom for each room!' (the exclamation mark is Kira's) (Kira, 1977: 8). Cooper points out that official winter-heating standards in the USA have risen by over 6°C over a 35-year period, in cases, requiring the renewal of equipment (Cooper, 1982: 243, 269, 288). These data must be put alongside the contemporary practice of marketing two-basin bathrooms ('his' and 'hers'). Given the fact that different cultures may well have different environmental expectations (Rapoport and Watson, 1972), much more needs to be known about how such expectations come about. (For recent changes in the use of similar consumer products in India, see King, 1989c.)

What, in other words, needs to be asked is whether there is an inherent logic in modern capitalist forms of urban and building development where there is a free market in property and land, which leads to Harvey's oft-quoted statement:

What is remarkable is not that urbanism is so different but that it is so similar in all metropolitan centres of the world in spite of significant differences in social policy, cultural tradition, administrative and political arrangements, institutions and laws and so on.

(Harvey, 1973: 278)

The reason, according to Harvey, is that the conditions in the economic base of capitalist society, 'together with its associated technology put an unmistakable stamp upon the qualitative attributes of urbanism in all economically advanced capitalist nations' (Harvey, 1973). The terms 'Western' and 'modern' blur the process taking place.

For example, in his study of traditional compound housing in three African cities, Schwerdtfeger concludes that both Islam and 'Western culture and technology . . . may well effect the gradual disappearance of large compounds and the increase of smaller, self-contained houses occupied by individual families'. Elsewhere, he writes of 'cultural change resulting in the reduction in size of co-residential families, the change in people's attitudes and consequently, the modification of traditional house-styles or the outright adoption of foreign patterns of housing' (Schwerdtfeger, 1982: 309, 312). Such 'Western' cultural changes seem very similar to the physical separation of work and residence and the development of consumer-oriented suburbs characteristic of capitalist forms of urbanization.

Similarly, criticism can be made about the use of the term 'international', a word which has both a general and, in the term 'international style', a more specific architectural meaning. As has been discussed in the previous chapter, the main period for the development of so-called 'international institutions' as well as the term 'internationalize' was in the half century before 1914, that is, at the height of European imperialism: the 'international powers' were, in effect, the colonial powers. The phrase, 'international style', as applied to urban and architectural form, was coined in the early 1930s to refer to the built forms of the 'modern', industrial, free-market society when world-economic and political relations were dominated by capitalist institutions; what is referred to by that term are the high-technology, high-energy materials and capital-intensive architecture containing the institutions of monopoly capitalism (see Hitchcock and Johnson, 1966). Likewise, the term 'European' as applied to built environments in the Third World; here it is worth quoting Wallerstein who writes that 'the expansion of the capitalist mode of production' could be thought of, quite misleadingly, as 'the expansion of Europe' (Wallerstein, 1976: 30). There is, in this context, however, still the need to distinguish cultural expressions of this mode, whether French, Spanish, British, or Dutch.

The emphasis in the preceding comments has been on global economic processes with a crude use of such notions as the 'capitalist world economy', a concept that refers both to a system of production as well as an ideology. Whilst technology (hardware) has the potential of being universal or international, ideology (software) is much less easily transferred, as recent history in the Middle East has shown. It is presumably at the level of ideology, understood in terms of beliefs, values, cultures, and the role of the State, that the study of social change should take place.

In terms of the analysis of built form, the overall theme of this chapter, a theoretical framework might be provided by combining two levels of analysis. On the one hand, are the broad generalizations of the urban political economy or structuralist approach that, in Harvey's (1973) terms, refer to 'the economic base of capitalist society with its associated technology'. If, as has been suggested on p. 93, all systems of economic organization or modes of production are expressed both spatially and in particular types of built form, then the validity of such concepts as 'the international division of labour' ought to be capable of being tested by reference to empirical evidence on the ground. This may refer, crudely, to types and forms of urbanization in different places and different periods, or, in a more sophisticated way, to the forms and functions of particular cities, for example, those that act as centres for headquarters of multinational corporations and, hypothetically, for the organization of production on a global scale (Cohen, 1981).

On the other hand, however, there is need of detailed empirical studies, both historical and contemporary, of particular local, regional, and national case studies that should demonstrate the degree to which local societies or cultures maintain control over and modify their own economy and polity, and hence their distinctive building and urban forms, in the face of global forces. In short, we need a variety of labels — social, cultural, and geographical, as well as those drawn from the vocabulary of political economy.

Chapter Six

CHANGES IN THE CORE (1)
The global production of building form

INTRODUCTION

In the previous chapters, I have suggested that much of the built environment, as well as other aspects of urbanism, both in core and periphery, has developed in the last two centuries — and longer in certain cases — as the product of an increasingly global system of production and international division of labour. What this chapter undertakes is to provide some empirical evidence for this proposition.

In particular, it discusses the 'theoretical fit' of particular concepts and paradigms, some of them introduced in the earlier chapters, for understanding and interpreting empirical data on one particular and specialized building form. Hence, the more general realm of enquiry is best conveyed in the form of three questions:

1. How is the built environment socially produced? In this context, the term 'built environment' refers specifically to building type (for example, factory, farm, prison, or church) or different types of dwelling (row houses, studio apartments, villas, or high-rise flats), the physical-spatial form or 'design' of these as well as the larger built environment (city, town, conurbation, as well as the rest of the socially constructed environment). The phrase 'socially produced' subsumes economic, technological, social, political, and cultural influences and determinants.
2. What function, purpose, and meaning does the built environment perform, and what implications does it have for the maintenance and reproduction of the larger economy, society, polity, and culture?

3. What theoretical paradigms are available for understanding the production of building form on a global scale?

Two common approaches to studying the production of the built environment can be identified, albeit in relatively crude terms:

(a) The historical, or diachronic, approach where the built environment is viewed as a product of social change in one society; this approach may also be seen as comparative with the comparisons made within one society over time.

(b) The cross-national and cross-cultural, or synchronic, approach where the comparisons are made between societies or cultures. Such comparative studies may focus on different levels of comparison, each of which, in its own way, has its own validity, though only yielding certain levels of explanation. For example, a study may be intrasocietal (the production of municipal housing in Manchester, Birmingham, and Glasgow in the twentieth century); intersocietal (a comparative study of State housing policies in France, the Netherlands, and the United Kingdom); it may focus on differences in political economy or political formation (housing policies in socialist and free-market economies), or on different cultures and regions (household and dwelling forms in the Apennines, the Pyrenees, and the Himalayas).

In the historical approach, the analysis is obviously not simply temporal or chronological; analytical categories relating, for example, to the system of economic organization or mode of production must be adopted, such as 'preindustrial, industrial, postindustrial', or 'feudal, mercantile capitalist, monopoly capitalist', which both structure the type of empirical data collected and also influence the kind of explanations offered. However, as so-called 'peasant', 'traditional', 'preindustrial', and 'late-capitalist' environments exist simultaneously in the contemporary world, an initial problem to be solved is in the labels and categories chosen to conceptualize the built environment on a global scale. One strength of the historical approach is the attention that it gives to specifically situational explanations — factors that are unlikely to reoccur in the future or at other times or places. Historical approaches can also take account of cultural variables without necessarily being 'cross-cultural'.

In the second approach, comparative or cross-national — cross-cultural studies focus, by definition, either on whole societies or on whole cultures, the latter defined in various ways but frequently referring to values, institutions, beliefs, and cultural specifics. 'Culture' has frequently been treated as an independent variable to 'explain' building form and the wider built environment (for example, Rapoport, 1969; 1976), particularly in cultural geography. Apart from this all-encompassing or 'anthropological' view of culture, however, the term may also be used in a partial way to distinguish certain levels of explanation; that is, the cultural may be separated from the economic or political. Furthermore, 'culture' may be used in a cross-cultural situation to refer to industrial culture or the culture of industrial capitalism, which may take similar forms in quite different cultural settings.

Irrespective of the validity of culture as an explanatory variable (discussed on p. 108), 'cultural explanations' pose methodological problems for empirical research: for example, how are values identified and demonstrated, and what models of society are assumed? Even without addressing the base-superstructure issue, other criticisms can be made. Thus, cross-cultural studies, by focusing on cultural variation, tend to 'chart' independent cultures: they become a self-fulfilling prophecy. Moreover, as is also the case with cross-national or intersocietal comparative studies, in focusing on geographically or historically defined societies, data collected and organized according to national or State criteria tend to be compared as such. The international flow of labour, capital, images, or ideas, which would illustrate processes of a different kind, tends to be neglected.

The theoretical shift associated with the 'new urban studies' has principally been applied to the study of cities — traditionally understood — examining their larger economic, social, and political context and exploring the critical role of the economic system in shaping the nature of urban and regional systems. In this context, the city is seen as a place of production and consumption frequently equated with capitalism, with attention being devoted to different levels of explanation with regard, for example, to the mode of production, the role of the state, etc. (see Saunders, 1981, for an early critique; also Smith, 1983; Walton, 1979).

Where the new urban studies have added considerably to the theoretical and conceptual apparatus of urban research (for example, in theorizing the state, notions of collective consumption, or the territorial and spatial division of labour (Massey, 1979; 1984)), they have, with some few exceptions, had limitations. Thus, with the

exception of Harvey (1973; 1985) or the work of the Bartlett Summer
School (1979–present) and until the mid-1980s interest in the 'post-
modern' (which is confined to the analysis of contemporary, rather
than historic forms), relatively little attention has been paid to the built
environment as a physical and spatial reality; the notion of 'space'
frequently utilized in this literature is rarely adequately spelled out
(for example, Castells, 1977; Pickvance, 1976; Saunders, 1981 (1984);
Gregory and Urry, 1985) and the actual tangible buildings and
bounded spaces are somehow etherealized or, at best, transformed into
abstract 'scarce resources', consumption items, or similar. The empirical base for much of the research has, till recently, been confined
mainly to the free-market and/or socialist societies of Europe, North
America, Australia, and, to a lesser extent, South America: Asia,
Africa, and the Middle and Far East have so far, and in comparison,
hardly been incorporated into this framework. And though empirical
historical research has been conducted, it represents proportionally
much less than theoretical and contemporary work. Nevertheless, some
of the perspectives developed in this body of research have been drawn
on in the discussion of the following issues.

EMPIRICAL RESEARCH

Within the general context of the two questions stated at the outset
of this chapter, the research in question has been concerned with an
attempt, first to describe and subsequently to explain, the meaning
and significance of one particular item (or segment) in the built
environment, namely, the specialized dwelling form of the bungalow.
Though an apparently insignificant, even trivial, example of built form,
the first relevance of the bungalow is that it is simply one among many
of a large variety of specialized building types: it is specialized in terms
of building form, location, function, construction, social and temporal
use, and historical specificity (that is, in terms of the particular
historical conjuncture of its introduction, or insertion, into the built
environment of the urban forms of industrial capitalism, and, subsequently, in terms of the numbers built as a proportion of the housing
stock, not only in the United Kingdom, the USA, or Canada, but also
in many other free-market societies). Hence, for the purpose of this
research field, it would be irrelevant *which* contemporary or historical
building type was chosen as the research object: the stock exchange, the
prison, the hospital, the crematorium, or the office block would be

equally valid. Within a general discussion on the sociology of the built environment, the relevance is that, by focusing on any one socially specialized and terminologically distinguished building type, questions are thereby posed about other types and, hence, about forms of social organization and their relation to the political economy or mode of production.

In this framework, therefore, social organization can be examined in relation to different types of building (and settlement): that is, the organization of people and social groups between offices, factories, schools, clubs, hotels, palaces; in addition, social groups, relations, and activities can be examined within different types of building (that is, between different rooms, spaces, locations, or floor levels). In short, the question to be addressed is: how are societies economically, socially, politically, and then spatially organized, not simply between 'urban and rural' or 'town and country' (crude and inadequate categorizations), or between functionally differentiated urban hierarchies, but within particular types of physically and spatially differentiated, spatially located built forms that themselves are terminologically distinguished and in relation to economic organization, increasingly on a global basis? Though a large number of (generally architectural) studies have been made of individual building types and their historical development (for instance, the factory, the workhouse, the hospital, the church, the school, or the bourgeois or working-class house), it is doubtful whether an institution such as the prison or the school can be understood adequately, either as a social institution or as an architectural form, without also considering its role and function within a larger theoretical framework that explains how the society as a whole exists and reproduces itself.

Within the comparative and historical mode of research that this chapter assumes, the identification of 'firsts', therefore, becomes important (that is, the establishment of the first purpose-built labour exchange, stock exchange, or treatment centre for psychiatric patients) as evidence of a change in societal arrangements by which particular tasks or 'functions' are accommodated. The inverted commas here are clearly necessary in that the whole question of a functionalist approach, which this may seem to suggest, with its assumptions about 'human needs', is, of course, tendentious. Some reference will be made to this issue later in the discussion.

In this context, therefore, the specific choice of the bungalow as an object of research was, in some respects, accidental. Earlier research on the social, political, and cultural context of colonial urban development had suggested the potential of the colonial bungalow as a peculiarly culture-specific dwelling form, associated with a particular mode of production, as an object of research (King, 1976: 13–15); undertaking a specifically historical study arose as a result of identifying what subsequently turned out to be the first bungalows to be built and named as such in the United Kingdom and almost certainly in the Western hemisphere (a relatively rare case of being able to identify the circumstances surrounding the production, or 'invention', in 1869, of what was to become a common building type). Though beginning initially as a semi-serious study in social history, the research subsequently turned out to have the potential for more serious enquiry: there were two reasons for this. In the first place, the bungalow is probably the only type of dwelling that, in both name and form, exists in every continent; such a phenomenon requires explanation. Second, though the initial research objective was to produce a straightforward 'cultural history' (from 1600 to 1980, tracing the bungalow from India to the United Kingdom, North America, Africa, Australia, and continental Europe), the investigation of that history raises a large number of important theoretical and methodological issues within the realm of the sociology of the built environment. The remainder of the chapter is addressed to some of these.

DATA COLLECTION

For the more limited purpose of this chapter, the term 'bungalow' is understood in its conventional twentieth-century European sense: in general, this refers to a separate, free-standing, or detached form of dwelling, principally of one storey, and located on its own plot. Not included in the definition is the fact that it is frequently or generally occupied by one household or nuclear family, or one generation of it.

Empirical data for the research have come from a variety of sources. On the assumption that socially and culturally significant phenomena are generally named, any reference to the bungalow as a terminologically distinguished dwelling type has been considered; all lexical data on 'bungalow' have been noted. This includes material from dictionaries, encyclopaedias, reference books, library catalogues, bibliographies, and specialized and general literature guides. For the early (Indian) phase, evidence comes from travel literature, ethnographic accounts, manuscripts, diaries, official records, and papers. Graphic, photographic, and cartographic records of the British in India have been consulted at the India Office Library, London, and, for the British in Africa, at other colonial repositories elsewhere.

For data on the introduction of the idea of the bungalow from India into the United Kingdom, North America, and continental Europe, journals, periodicals, and books on building, architecture, planning, housing, homemaking, etc., have been consulted. The *Avery Index of Architectural Periodicals* provides references to the main European (for example, German, Dutch, Italian, French, etc.) articles. Other data have been located in newspapers, builders' catalogues, postcards, poetry, novels, and sheet and recorded (1920–1980) music. Contemporary statistical data on housing and dwelling stock are available from official government sources and building-society records. In addition, secondary sources exist in the form of books, articles, and theses, mainly from the USA and, to a lesser extent, from Canada and Australia. Almost all of these data, however, are in the form of architecture or architectural history and require 'recycling' to address the issues listed as follows.

Fieldwork, of a relatively unsystematic nature involving my own observations, was undertaken during a residence of five years in India and on visits to the USA and Canada and intermittently, to parts of Europe as well as in the UK. Other information on the existence and use, both of the term 'bungalow' and of the form to which it applies, has been gathered from students from different parts of the world studying at the Development Planning Unit, University College, London.

'Making sense' of these data and structuring them into some meaningful form has required drawing on perspectives and theory from three fields of research:

1. the social production of building types (for example, prisons or asylums);
2. the political economy of urbanization; and
3. urbanization and the world-economy.

Other theoretical fields relevant to the interpretation of the data (for example, with regard to cultural production and reproduction) are not discussed in this chapter.

In the next section, basic methodological questions presented by the empirical data are stated; these questions are conceptualized, or theorized, in the headings immediately above them. Though space does not permit the discussion of all the issues presented, brief comments are made in regard to some of the theory drawn on, its adequacy, and, where appropriate, its shortcomings. It will be obvious that the range of issues raised, combined with the amount of data collected, prohibits any detailed (let alone comprehensive) coverage of each of the theoretical fields referred to. What has been attempted in the study is a (lightly) theorized history in which only some of the theoretical sources are referred to. The questions posed in the next section are in the order suggested by the historical treatment of the topic rather than in what might be suggested by a more ordered 'model'.

THEORETICAL APPROACHES

Economy, society, and culture and their relation to building form

How do we explain the 'Anglo-Indian' bungalow as a distinctive and specialized form of dwelling in colonial India?

The production of the Anglo-Indian bungalow in India in the eighteenth and nineteenth centuries poses questions of explanation at different levels. In the first place, the political economy of colonialism focused attention on the unequal distribution of power, between Europeans as colonizers and Indians as colonized, in the production of the built environment: a situation in which a cultural division of labour needed to be built into any explanatory model. The initial approach to this problem was, therefore, to adopt a straightforward 'culturalist' position, using work in social anthropology but overlaid by emerging theory on dependency and on notions of dominance — dependency, dominance, and conceptualizations from world-systems theory on 'core-periphery' (King, 1976). Implicit in this approach were assumptions about 'cultural autonomy' understood in terms of values, beliefs, ideas, and modes of representation.

Explanations, therefore, placed emphasis on sociocultural variables (that is, in relation to the cultural use of space, or to household and family structure). These explanations were suggested by the obvious cross-cultural context in which European social forms (and built forms in which they were contained) were compared with indigenous Indian forms. Theoretical insights were taken from varied sources in social

107

anthropology (Hallowell, 1967; Rapoport, 1969; and others) and from work on cultural pluralism (Smith, 1965). Subsequently, a growing literature both in architecture and in anthropology has focused on the social and cultural variables in the use of space, on social and spatial symbolism and organization (for instance, Ardener, 1981; Douglas, 1979; Doxtater, 1984; Carswell and Saile, 1986; etc.).

This approach has been (rightly) criticized as overly 'culturalogical', giving inadequate attention to the economic history of the context in which the processes took place (Gutkind, 1981). The culturalist position tends to be static, giving too little importance to the dynamics of economic, social, and cultural change.

In short, the issue on which this question focuses attention is the transformation of space usage, including the specialization of rooms and changes in household form in the metropolitan society consequent upon the emergence of industrial capitalism in the eighteenth and nineteenth centuries, in comparison to indigenous 'pre-capitalist' Indian forms (King, 1984). The question of how and why domestic space (the size, form, shape, location, and variety of dwellings) developed as it did in any particular society (in this case, the United Kingdom) during the processes of capitalist industrialization, both *before* and *during* the period in which such practices were transferred to India or Africa through colonialism (or to North America and Australia), needs to be more carefully addressed. Marxist work on the production of space under capitalistm (Lefebvre, 1970, referred to in Saunders, 1981; and Lojkine, 1976), particularly in relation to the construction and leisure industries (also see Johnston, 1980), is suggestive, though it also tends to be opaque and indirect. Work by Ball (1983: 136–43) lends a greater degree of specificity to the way in which actual physical and spatial forms of housing are produced by the system of economic organization and housing production. The basic problem, however, is to explain how the particular physical and spatial forms of the built environment, including a given set of building types, emerged as it did, and what the 'cultural variable' (the English, Dutch, German, or French) of this is. Some insights into this question may be gained from studies of culture and ideology (for example, the work of Adorno and Horkheimer (see Held, 1980)). Related to these issues is the next question.

Culture and the political economy of building form

How do we explain changes from indigenous dwelling forms (whether in India or Africa) to 'European' or 'modern' forms?

Subsumed under this heading is the question of the transformation of material culture, particularly domestic space and buildings in 'non-Western' peasant societies; it raises issues about colonialism in the development of capitalism as well as about the development of capitalist modes of production, both in different geographical areas as well as on a global scale. In historical terms, it concerns various aspects of the transition from feudalism to capitalism; for example, the change from subsistence agriculture to the production of cash crops, the nature and causes of urban development, the transformation of the building process, the introduction of new materials, the transplantation of buildings and consumption habits, the importation of consumer goods, etc. Also in this context are questions concerning the meaning and significance of norms of building and spatial design, and their transfer (both historically and currently) between the core and periphery, as discussed in the previous chapter.

In everyday language, these issues are subsumed under discussions of 'Westernization', 'Europeanization', 'modernization', or 'bourgeoisification', each term carrying different connotations. Although the literature on these issues is vast (some of it usefully discussed in Oxborough, 1979), theoretical and empirical studies provide interpretative frameworks. Mabogunje (1980), for example, discusses the creation of markets, the introduction of wage labour, and the development of capitalist forms of urbanization in relation to Africa. Though studies dealing directly with the role played by the built environment in the process of economic, social, and cultural transformation are few, *aspects* of these processes have been examined. For example, the introduction of capitalist notions of property are discussed in relation to North Africa, West Africa, and India in Abu-Lughod (1976; 1980), Hopkins (1980), Neild (1979), and Bayly (1983) and the development of a real-estate market, in the context of shifts in the world economy, in Ribeiro (1989). Alavi (1980) provides a valuable theoretical and empirical discussion relating to India that covers the impact of colonialism on landholding and the introduction of metropolitan notions of property; the transformation of relations of production and the creation of bourgeois landed property; the significance of colonial land revenue for capital accumulation; the effect of a developing market economy, foreign trade, and land revenue on urban growth, and on the development of a new Indian bourgeoisie.

These theoretical themes have been fleshed out by Bayly (1983), who traces the construction of urban property by landed proprietors and the adoption of 'Western' goods in what he refers to as 'the colonization of taste'. Alavi (1980: 387) quotes Gadgil (1973) on the newly created Indian bourgeoisie, which 'showed itself ready to accept European standards and pour scorn on everything Indian'. Other studies chart the changes in taste and cultural criteria of perception that accompany these economic processes (Tarapor, 1984; see also King, 1984; Oldenburg, 1984).

In historical studies, the direct linking of these processes to questions of commodification, however, is rare, although developments clearly have importance in terms of changes in consciousness, ideology, social relations, and class structure. There is some discussion of the role of transnational corporations in changing cultures and creating markets for Western goods in Kumar (1980), and attention is given to life-styles and suburban modes of residence of 'international elites' in Sunkel and Fuenzalida (1979).

With regard to contemporary, as opposed to historical, developments, Burgess's (1978) theoretical critique of the writings of Turner provides insights into emergent housing processes in the transition to capitalism and the characteristics of land and housing in a capitalist economy, including the transformation of self-help housing into commodity form. Burgess also discusses the relations between land, private property, and the law, as well as questions of minimum standards and prescriptive legislation. Both Kanyeihamba (1980) and McCauslen (1981) also deal with the relation of regulations and legislation to the production of housing norms and, in Burgess's words, 'the ideology that governs the middle class image of the finished housing object and the schools of design whose purpose is to rationalise the needs of this form of housing production' (Burgess, 1978; 1124).

The social production of specialized dwelling and building forms

How do we explain the introduction of the bungalow, as a specialized 'vacation' or 'holiday' dwelling in the United Kingdom in the third quarter of the nineteenth century?

In the case in question, this presupposes the production of the resort as a distinctive form of urban settlement; this raises questions about,

110

for example, the relationship of urban development to changes in the mode of production (industrial capitalism); the creation, disposal, and investment of surplus, and the need to examine these processes during the emergence of a global economy and the developing international division of labour. It also poses questions about social organization and its relation to spatial organization, as well as the social organization of time (King, 1980).

The role of the city as a place of production and consumption is central to the theoretical concerns of urban political economy and the differentiation and segmentation of markets is also a standard theme in economic theory. With regard to the specific location of the bungalow as specialized vacation house within commuting distance of London, the work of Massey (1979; 1984) and others on the developing spatial division of labour in the nineteenth century and the emergence of specialized areas of industrial production, investment, and consumption is suggestive. The expansion of leisure, entertainment, and recreation under late-capitalism and its relation to developments in the built environment are discussed in Johnston (1980: 118). Particularly suggestive is Harvey's (1985) essay, 'The urban process under capitalism', on the accumulation and circulation of capital in the built environment, as well as his initial work in this sphere (Harvey, 1973).

So far, however, the object of research in this field has been 'the city', 'the urban', or 'the built environment' in general, and has not been related to the production or creation of new building types or forms; yet it is clear from the historical evolution of different types of building that in the period of the greatest growth of industrial capitalism between 1850 and 1875 (Hobsbawm, 1975), a large number of new building types were either developed or created, as discussed in the previous chapter and that, currently, some of the largest developments in building and architecture are in the so-called leisure industry, which functions to keep capital in circulation and to promote consumption. (*The Times* (1984a) reports on the launching of a company 'to invest millions of pounds in the leisure industry to help cope with the expected increase in free time'. It expects to invest in health hydros, country clubs, fitness centres and dance studios, hotels, holiday villages, marinas and 'theme' restaurants.) Other 'new' or modified building types that might be mentioned in this context are leisure centres, theme parks, shopping malls, and indoor stadia. Current political developments in the United Kingdom provide a rich research arena for examining changes in political economy and their effects on building form: the physical and spatial results of transferring responsibility for housing elderly people from public to private sectors, the promotion of private health-care facilities, or the disposal of nurses' hostel accommodation as part of health privatization policies.

The full development of the 'holiday' resort in the second half of the nineteenth century was also the product of new space-time relationships consequent upon the emergence of industrial capitalism; the emergence of the middle-class 'week-end' [sic] (1880–1900) was a prerequisite for the production of the 'week-end [sic] bungalow' or 'cottage' (the earliest reference to which is in 1904), a specialized built form embodying a temporal-spatial unit (King, 1980). Temporal and spatial organization is discussed extensively in Carlstein et al. (1978), and Giddens (1984). Duclos (1981) has also provided useful insights into the contemporary global system of the management of time in capitalist societies.

The relationship between forms of economic and social organization and between specialized building and housing types

How do we explain the increased specialization of building types in industrial and in industrial capitalist societies?

This issue is related to those discussed in the previous subsections. In addition, however, it poses questions about the categorization of dwelling forms and their relation to economic and social organization. For example, what economic or social significance is attached to the standard categorization of dwelling forms in contemporary Britain (detached, semi-detached, and terraced houses; purpose-built and non-purpose-built flats; bungalows), as well as to the further development of subcategories of these? Some insights are provided by research on increased consumption represented by separate dwellings in scattered settlements (discussed in relation to tenure and consumption; p. 113). On a more local level, Ball (1983: 139–41) has highlighted the way in which housing developments are segregated into market sectors with particular consideration being given to detached houses as a way of maximizing profit.

The 'second home' in relation to economic or urban theory

What is the meaning and significance of the historical development of the bungalow as a 'second home'?

Much research on second homes (up to 1980) has tended to be empirical and concerned either with questions of policy ('leisure planning') or of social equity; the theoretical conceptualization of second homes is relatively undeveloped (also see King, 1980), with little attempt so far to link this question with temporal or spatial aspects of urbanization. Here again, Harvey's (1985) framework on the accumulation and circulation of capital in the built environment provides the most satisfactory framework for the discussion of second homes, particularly in the context of the recent extensive growth in time-sharing and multiownership.

Relationships between per capita income, forms of tenure, and dwelling form in different societies

How do we explain the increase in the number of bungalows as a proportion of the total housing stock?

This again may be seen as a subtheme of the larger topic outlined in the first section on p. 108. It raises questions about, for example, the extent of economic development in a society, industrialization, political formation (capitalism in various forms, socialism, etc.), geography, and a variety of culture-specific and historically specific explanations.

Work by Duncan (1981), Agnew (1981), and Kemeny (1981), undertaken across a number of different free-market economies, has attempted to relate (a) per capita income in different societies to the proportion of households in different forms of tenure (that is, mainly public and private renting, owner-occupation); (b) per capita income to proportions in owner-occupation; and (c) per capita income, proportions in various forms of tenure, and proportions in different forms of dwelling (that is, multifamily dwelling forms, such as apartment houses, and single-family detached houses). Though societies with high rates of per capita income also have high rates of single-family dwellings (for example, the USA and Australia), especially if compared to the predominance in cities in the USSR of multifamily, multistorey forms, other geographical, historical, and cultural variables cannot be eliminated; hence the conclusions reached by these three scholars are tentative. In addition, considerable contrary evidence is quoted in these studies. For example, Sweden, West Germany, and Switzerland all have high per capita incomes, though with different proportions in different tenures and relatively high proportions

113

of households in multihousehold forms of dwelling.

In this context, the historical study of the bungalow reinforces culture-specific and historically specific, or situational, explanations for developments in the built environment, though, as this building form is also becoming especially characteristic of free-market economies, some generalizations along political-economy lines seem legitimate.

The relationship between dwelling form, urban form, and settlement form

What is the significance of the bungalow in the historical development of outer suburbanization, especially in the USA where it is the fore-runner of the contemporary 'ranch house'?

In discussions about the production and consumption functions of 'the urban', and especially, 'the suburban', suburbanization has been seen as a form of built environment both characteristic of, and intrinsic-ally produced by, industrial-capitalist forms of urban development. Yet little analysis has so far been devoted to differentiating between different *forms* of dwelling in suburban developments; that is, row housing, multistorey apartments, or the single-storey, single-family, detached bungalow or ranch house.

Suburbanization has been the object of considerable theoretical debate in sociology for at least three decades. The literature is exten-sive and includes a large number of studies that have attempted to relate a particular physical and spatial urban form (the suburb) to particular kinds of social behaviour (suburban life, suburbia). Subse-quently, attention has been focused on the relation of suburbaniza-tion to consumption in market economies (Harvey, 1973; Walker, 1978; 1981; Lash and Urry, 1987) with emphasis given to various in-terests, ranging from property developers, estate agents, automobile manufacturers, tyre companies, surveyors, highway engineers, super-markets, and producers of consumer goods interested in suburban expansion and 'sprawl' (for instance, Checkoway, 1980). Surprisingly little attention, however, has been paid in this literature to the actual *form of dwelling* in suburban development and to the ideology that legitimizes it. Whether suburban residents live in fifteen-storey high-rise apartments, without yards, gardens, or swimming pools, or the ability to exercise control over the external appearance of their dwell-ing, or whether they live in split-level detached ranch houses with sprawling yards and/or gardens, is obviously related both to patterns

114

of consumption as well as to social relations. Symbolic elements are also of importance in relation to the representation of identities.

Though Tabb and Sawers (1978) refer to architectural structures that manifest a 'consumerist orientation' in market economies, the simple but radical historical transformation whereby dwelling forms changed from typically two- and three-storey forms to those of a single storey is obviously of considerable importance in this context.

The building process, prefabrication, and urbanization

What is the significance of the bungalow as the archetypical product of prefabricated building techniques?

The (often) self-built bungalow developed in the twentieth century frequently resulted from the utilization of new industrialized building materials (asbestos, corrugated iron, rubberized roofing materials, machined timber), many originating in the context of an emerging world-economy and new modes of transport.

The theoretical issues raised here relate to questions of deskilling and the marginalization of labour, to access to property ownership, and to control over building development. Some of these topics, including the question of monopoly control over building materials, are discussed in Ball (1983), and, as in the section on 'culture and the political economy of building form', pp. 109–10, in Burgess (1978). The issue of building development control is discussed in the next section.

The social control of the built environment and the function of the environment in social control and in the reproduction of society

How do we explain the controversy over the location of the bungalow in the United Kingdom and the response in 'country planning' legislation?

A variety of conditions, combined with those indicated in the previous section, led historically in the United Kingdom to building development in what were culturally defined as 'country' areas, previously the territorial preserve of social elites; the response to this 'invasion of the countryside' in the interwar period was the introduction of 'town and country planning' legislation enforcing particular divisions, which are simultaneously economic, social, cultural, and political, on urban and rural areas (for instance, the 'Green Belt') (Hall *et al.*, 1973).

This issue raises a variety of theoretical problems concerning the social control of the built environment, the reproduction of environments, and the reproduction of societies (that is, what function does the maintenance of 'town–country' divisions, through 'planning', have in relation to the reproduction of social divisions?). Although some aspects of these questions have been discussed in theoretical debates on community power and the sociology of planning, the residual or inertial effect of the built environment on the reproduction of social formations (and this does not imply either environmental or architectural determinism) remains relatively unexplored, though Foucault's work is obviously pertinent. As Gershuny (1978: 7) points out, within given structures of class relationships, 'cultural forms may have an inherent tendency to replicate themselves over time'.

Urban and building form and the world system

What is the significance of the introduction of the bungalow as a specialized form of outer suburban or country dwelling at particular historical conjunctures, 1890–1910 and 1920–40, in the United Kingdom?

As a specialized dwelling type, distinguished by name and single-storey form, the bungalow was introduced into Europe at a particular place and at a particular time: this was not in France or Germany in the mid-eighteenth or the late twentieth century, but in England, especially between 1890 and 1910. An explanation of this is required that is more convincing than a simple notion of 'diffusion'. In this context, the particular form of the bungalow is representative, both as product and producer, of various larger urban phenomena: outer suburbanization, resort development, and the introduction of purpose-built 'second homes', and (as 'retirement bungalow') household fission; all these phenomena may be related.

The theoretical issue here, therefore, already touched on on p. 109, concerns the relation of culture and ideology to political economy: how is culture (in this case, not only urban and architectural form but the ideology and life-styles that they represent and contain) produced by particular economic and social relations?

The most plausible explanations for the appearance of the bungalow at particular conjunctures in history derive from world-systems theory, particularly work on the world-economy and the international division of labour. The industrial predominance of the United Kingdom in the nineteenth century was part of an emerging global system of production, of which another part was the colonial restructuring of once self-sufficient peasant economies in Africa and Asia towards cash-crop production and monoculture plantation agriculture (Hobsbawm, 1975). The development of industrial urbanization in the United Kingdom was likewise accompanied by, and linked to, agricultural development in Latin America and the commercial orientation of Latin American cities (Browning and Roberts, 1980; Roberts, 1978). In an increasingly sophisticated world division of labour, London was the largest world city, whose commercial and service-oriented population (many people were directly or indirectly employed in relation to the imperial economy) encouraged extensive suburbanization in the late-nineteenth and early-twentieth centuries (Jackson, 1973: King, 1989).

The decline of agriculture (and agricultural land values) caused by the emergence of world markets in agriculture took place at the same time as the extensive accumulation of capital generated by London's increasingly specialized function as world finance and commerce centre (Ingham, 1984). A combination of these developments was behind the boom in country house building in and around the London region and the introduction of the first middle-class 'country bungalows' during the period 1890–1910 (Aslet, 1982; King, 1984).

The first expression of the bungalow in the Western hemisphere — in the urban environments of industrial capitalism — was technically from 1869 to the mid-1880s and on the Kent coast (King, 1984, Chapter 2). Yet in one sense, this was, in terms of 'built form', possibly a 'false start', as the historical phenomenon was more in relation to an idea, a definition, and a term, rather than the 'pure' reproduction of the form itself. And the numerical significance of the phenomenon was very small (less than ten examples known, and all in one place, before 1884). Despite this, the conditions giving rise to its production — the investment of surplus acquired on a global scale, the invest-ment in consumption-oriented urban forms, and — most significantly — the spatial location of these activities (and forms) at the margins (or outer peripheries) of London, as the major world centre for capital accumulation, during *the* major decades of capital accumulation up to that time — should be noted. Yet incontrovertible evidence exists to demonstrate that the principal introduction of the bungalow form (and the associated cultural practices) takes place between about 1890 and 1910. Why this particular time and place?

Prior to the late 1880s, the form and idea (as material cultural expression) was virtually unknown in the UK. Where it was known was in colonial Asia — India, Malaya, Ceylon. How was it known? Especially as the planter's bungalow, or in the hill station, the civil lines, or as the magistrate's or district commissioner's dwelling (Figure 6.1).

Basically, it was part of the built environment of a colonial political economy: the planter's bungalow, part of a system of cash-crop production operated by representatives of a particular culture in which local labour ('natives') lived in self-built huts and managers lived in an evolved, culture-specific dwelling form known as a 'bungalow' (King, 1984, Chapter 1).

Overall, this was a system of agricultural production (of tea, coffee, cocoa, rubber) as well as of politico-administrative control (civil station, railway network), and military power (cantonment), which amounted to a colonial mode of production (Alavi, 1980; Wolf, 1982, Chapters 3, 10). Societies where this took place were very little urbanized, primarily oriented to agriculture, where the growing towns were not industrial, but acted as port cities, administrative or commercial centres whose inhabitants either controlled the political-economic system or acted as consumers or as a labour force. The products of that colonial mode of production (supervised by planters and colonial officials) were increasingly available (as tea, sugar, coffee, cocoa, and cotton) to lower the costs of labour in British industrial cities, where labour lived in urban dwellings (often terraces or 'back-to-backs'), totally inhibited — both through lack of time, energy, and especially accessible land (agricultural space) — from any kind of self-provisioning.

The 'genuine bungalow' (in the form illustrated here) arrived in the United Kingdom from the 1890s, through a process of 'diffusion' but only under particular economic, political, and social conditions. By the nature of the form, and its spatial requirements, it can only be introduced as a cultural model of living in *non-urban*, or *ex-urban* areas. In the 1890s one could not, or would not, locate a bungalow in the middle of a city (ironically in the 1980s, 'user friendly' bungalow designs, for publicly-funded health centres, libraries, and community buildings, are built on city-centre lots vacated by capital flight): not only — because of the development of an international division of labour — were agricultural land values going down but, through Britain's predominant position in the world economy at that time, the economic benefits (the proportion of global surplus) of this were

118

Figure 6.1 India, 1860s (courtesy, India Office Library)

expanding the size of the urban bourgeois class. These, with a long-established model of 'country living' presented by the elites, were the people who took up the particular cultural model of the bungalow that colonialism, and the international division of labour, had made available. The evidence for these statements is, firstly, in *Bungalows and Country Residences* (Briggs 1891) (the first ever book on the bungalow as a 'modern' architectural form in the industrialized West), published between 1891 and 1901, and, secondly, in the journal *Building World* 1902 (Figures 6.2, 6.3, and 6.4). The publication of this first book of designs on this new form of 'ex-urban' dwelling and 'second home' coincided with a major boom in house-building in London (Parry, 1965: 208). The connection, and similarity in *form* (only) between these bungalows in Britain and those in the colonies is clearly direct. But of more importance are the complementary economic, political, and, not least, spatial conditions or structures that had produced simultaneously, and at both ends of a single spatial division of labour, this particular cultural phenomenon.

This 'true' phase of (highly select) bungalow development was for middle-class occupation in outer suburban, and especially country locations. Everything about their form, utilization, land use, and siting is evidence that they are essentially building forms geared to *consumption* — of space, views, scenery, time, consumer goods, and money — and this was consumption in agricultural areas, previously devoted to *production*. They were also, obviously, evidence of accumulation in that they were a second dwelling. Moreover, the 'spilling out', or expression of that surplus was, again, not in Scotland, Wales, or the Midlands, but on the reachable margins of the centre of capital accumulation — London. The later (*c.* 1900–10), socially 'downwards' diffusion of the bungalow was an expression of this increased consumption, in 'leisure', and a higher standard of living, which colonial expansion, and Britain's predominant position in the capitalist world-economy, had made possible.

If one line linking a spatial division pf labour can be drawn between the United Kingdom and India, another can be drawn from the UK to other parts of the globe — Latin America, West Africa, Malaya, and elsewhere. Here, industrialized capital-intensive production was being used to produce prefabricated dwelling forms (bungalows) to export to Chile, South Africa, the Argentine, and elsewhere on the periphery, again for the managers of mines, railways, plantations, managing British capital and exporting raw material

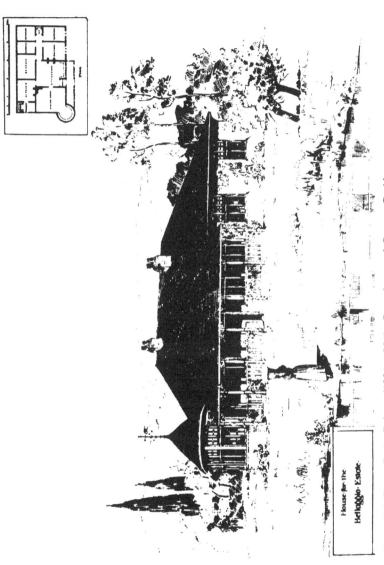

House for the
Hellingloo Estate

Figure 6.2 From R.A. Briggs (1891–1901), *Bungalows and Country Residences*

Figure 6.3 From R.A. Briggs (1891–1901), *Bungalows and Country Residences*

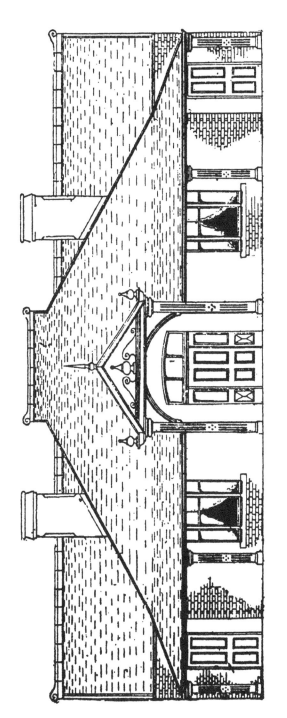

Figure 6.4 'This type of house is prevalent in India and most of our colonies, and is being increasingly adopted in the United Kingdom as a convenient and comfortable form of summer residence.' From *Building World*, 1902: 319

to industrial economies at the core (Figure 6.5).

In short, the bungalow is one item in a larger system of urban and architectural expression (urban planning is another; see Chapter 3) of a single division of labour characteristic of the capitalist world-economy.

With other economic, social, and political factors, similar structural explanations can be offered for the growth of bungalow developments as part of extensive suburbanization in the period between 1920 and 1940, particularly in south-east England ('bungaloid growth', a term coined in 1927, was the predecessor of 'suburban sprawl'). The collapse of staple export industries with world depression and competition from colonial and newly industrializing states (especially in textiles) between the wars led to massive industrial and urban decline in traditional industrial areas in the north-east and north-west of England. Sectoral and regional restructuring was to be responsible for new industrial growth in the South-East, with extensive suburban growth, again encouraged by the collapse of world-commodity prices in agriculture (Ball, 1983: 30; Carr, 1982).

The relationship between household fission, property ownership, and building and dwelling form

What is the significance of, and how do we explain, the 'retirement bungalow'?

This topic also covers a wide range of issues, concerning, for example, the expressive and symbolic function of dwelling form, market segmentation and the accumulation of capital, and the individualization of consumption. It also needs to be discussed in relation to changes in family organization and values, as well as in terms of the relation of the life-course to the mode of production.

In the capitalist economies of Europe and North America, separate accommodation for elderly people in the private sector is frequently represented by the detached, single-storey, 'retirement home' (in the United Kingdom, specifically referred to as the 'retirement bungalow'): since the early twentieth century, this has become increasingly institutionalized as part of larger residential areas in 'retirement resorts'. (A report sponsored by the National House Building Council (NHBC, 1984), a body supported by public and private sector organizations, concluded that there was a potential demand in the United Kingdom for between 250,000 and 400,000 units of 'sheltered retirement

BOULTON & PAUL, Ltd., Manufacturers, NORWICH.

IRON BUNGALOW.

No. 90/XY.

Several Buildings to this Plan have been erected in Portuguese West Africa, South Africa, Chili, and the Argentine.

Code Word—Timber.

PRICE - - **£430**

f.o.b. London.

Cast-iron Pile Foundation £40 extra.

Approximate weight 22 tons.

Approximate measurement 1800 cu. ft.

———*———

Cables:
BOULTON, NORWICH, ENGLAND.
A.B.C. Code used, 5th Edit.

Figure 6.5 Buildings and the international division of labour

accommodation' (a form of purpose-built housing for elderly people, grouping together bungalows or flats, with a warden or neighbour able to provide help). The research, based on a nationwide survey, found that two-bedroom bungalows were the most popular type of accommodation (*The Times*, 1984b).

Some of these issues have been discussed by Gershuny (1978: 6–7); among his conclusions on the post-industrial society is the notion of the self-service household, with increasing capital investment in the home and an ever-increasing proportion of final consumption taking place there. As the self-service household becomes more self-sufficient, it is inward turning and less sociable (p. 148).

Part of the explanation for this conclusion (with its implications for dwelling form) is what Gershuny calls 'inertial mechanisms which tend to limit the malleability of the future', in particular:

> the existence of a large amount of laid down capital in the form of a stock of dwellings which are designed for a single nuclear family, laid out in diffuse settlement pattern, places a limit on the possibility of radical change in the methods we use to meet our needs . . . the existence of capital in this form must tend to maintain us on our course of increasing consumerism.
>
> (Gershuny, 1978: 6)

In short, 'cultural forms may evolve so that they have an inherent tendency to replicate themselves over time' (1978: 7).

Counterurbanization and the capitalist world-economy

How do we explain the existence of the bungalow, both in name and in form, in North America, Australasia, and continental Europe?

As a specialized outer suburban single-storey dwelling form, used for permanent residence, the bungalow — both in name and in form — was introduced into North America in about 1905 (especially as the 'California bungalow') and into Australia shortly afterwards. From then until about 1930, it became a predominant and characteristic form of suburban dwelling in both continents. According to North American authorities (Winter, 1980; Lancaster, 1985), it was also the prototype of the contemporary 'ranch house'; as this single-storey prototype, it was also to help effect the transition from an earlier 'vertical', or 'two-storey', inner-suburb form (dependent on public transport), to the contemporary 'horizontal', single-storey, outer-suburb form (dependent on the private car). Preliminary evidence indicates that it was also introduced, again in name and in form, from about 1930 into France (Benoit-Levy, 1930) and from 1950, into the Netherlands, Germany, and Scandinavia (for example, Betting and Vriend, 1958).

In 1976, Berry suggested that 'counterurbanisation has replaced urbanisation as the dominant force shaping the nation's [that is, USA] settlement pattern': counterurbanization was defined as 'a process of population deconcentration implying a movement from a state of more concentration to a state of less concentration' (Berry, 1976: 17). Since then, other research has demonstrated that the process of slowing down in the growth of, or more frequently, the actual decline in, the size of major metropolitan regions, and the move of population back to rural areas and small towns, has been widespread in the major 'advanced' capitalist industrial states or, in the terms of Lash and Urry (1987), the areas of 'disorganised capitalism'.

Thus Vining and Kontuly (1978) showed that in Japan, Sweden, Norway, Italy, Denmark, New Zealand, Belgium, France, both Germanies, and the Netherlands, there was a reversal in the net population flow from their sparsely populated peripheral regions to their densely populated core regions, or a drastic reduction in the level of that flow. Fielding (1982), covering more or less the same countries in Western Europe, has also charted this process of counterurbanization: after considering the three explanations frequently put forward for the phenomenon (Berry's (1976) 'counterurbanisation model', the neoclassical economic model, and the State-intervention model), Fielding suggests that the most convincing explanation is to be found in the shared economic and social forms that they manifest as 'mature capitalist societies': similar social-class structures, similar public institutions, and a similar mass culture (Fielding, 1982: 25). Harvey (1973: 278) has also commented on the similarity of urban forms produced by advanced capitalism.

Certain aspects of Fielding's (1982) discussion are of central relevance to addressing the question posed in this subsection. The first relates to his consideration of migration and social class, where attention is drawn to the increase in migration of salaried functionaries whose residential preferences are for suburban and exurban environments. The second relates to his discussion of migration and the changing geography of production. Here he discusses the now-familiar theories associated with regional restructuring accompanying deindustrialization in the older industrial regions:

> From this perspective, counterurbanisation is seen as a product of the rapid deindustrialisation of the largest cities and old industrial regions (e.g. Ruhr, Nord, Merseyside) of Western Europe, accompanied by a stabilisation of rural population in small and medium sized towns in rural and peripheral regions.
> (Fielding, 1982: 31; also see Herington, 1984; Lash and Urry, 1987)

The adoption in Europe of the bungalow as an archetypical outer urban dwelling form may be understood in relation to these developments. Moreover, to use the industrial areas of the United Kingdom as an example, it can also be explained by the increasing emphasis on consumption in the economies of late-capitalism. In the nineteenth century, the typical form of working-class dwelling in the industrial cites of northern England was the terrace or, often, the minimal 'back-to-back' (Daunton, 1983; Muthesius, 1983). The logic of their design and location was to provide labour power close to the factory at minimal cost.

In the late twentieth century, in which labour has been increasingly dispensed with, either by technological developments or by the transfer of production to low-cost factories in the Third World, such minimal dwelling units have no logic. In the 'advanced', increasingly service-oriented economies of Europe, including the United Kingdom, there is increasing emphasis on consumption. For this, growing numbers of detached houses and bungalows in the outer suburbs provide the opportunity not only for one-car ownership, but often for two-car ownership, per household (increasing in the United Kingdom from 2 per cent to 15 per cent between 1961 and 1981 (Ball, 1983; King, 1984)).

The political economy of global urbanization

How do we explain the transformation of built environments on a global scale?

The appearance of the bungalow in its three main functional and locational contexts — as 'second (vacation) home', as suburban or exurban residence, or as privatized property for the socially 'retired' in many countries of the world in the late twentieth century is one of a number of examples of the global production of building form (another obvious example is the high-rise office block: together they provide the profile of the outer to the inner city). In a historical sense, it poses questions about the structures, mechanisms, and processes in the emergence of the capitalist world-economy (Wallerstein, 1979) and, more particularly, about the physical and spatial environment that characterizes that economy. It suggests the analysis of environments in relation not only to the mode of production but also to the territorial scale (local, regional, national, or global) on which that mode of production operates. Especially from the nineteenth century onwards, and in particular in the present day, that mode of production has been a capitalist world-economy (Wallerstein, 1974; 1984).

This concept itself drew on a range of earlier research on 'the development of underdevelopment' although cities and environments do not figure in this work. Of more importance in this context is the work of Braudel (1984; 1985) on the structures of everyday (material) life. Earlier chapters have referred to the growing body of work in the 1980s on world political-economy approaches to urbanization, and also concepts of dependent urbanism.

In this literature, however, little attention has so far been given to the way in which the built environment, either at the core or at the periphery, was affected by — or affected — the global relations of production: Roberts (1978), for example, concentrates mainly on economic, social, and political processes, and recent geographical studies conceptualize imperialism in terms of its having an 'impact on the landscape' (Christopher, 1988). This lack of attention to the role of space and built form in transforming economies, societies, and cultures is surprising since the urban, building, and architectural forms extending through the colonial urban system provide evidence, not only of the change from one mode of production to another, but of the emergence of such a global system of production (King, 1976; 1984). The transformation of the physical and spatial environment, especially in the form of cities, was a prerequisite, on one hand, for establishing 'Western' (that is capitalist) patterns of consumption and culture in the 'non-Western' world. On the other, it also provided possibilities for different kinds of consciousness for the representation of different kinds of cultural, ethnic, or national identities. The outcome is a mixture of both.

Chapter Seven

CHANGES IN THE CORE (2)

*Building, architecture, and the new
international division of labour*

A THEORETICAL FRAMEWORK

Whilst it is now fully recognized that contemporary urban change in
Britain results from changes in world-market conditions (e.g. Cooke,
1986; 1989; Massey and Meegan, 1982; Pahl, 1984; Rees and
Lambert, 1985), the earlier chapters of this book have argued that the
historical phenomenon of urbanization is still generally treated as a
nationally autonomous process. It is as though cities somehow
developed independently of the world outside Britain, of the sources
of raw materials that were the prerequisites for early and subsequent
industrialization, the overseas (often urban) markets for which urban
manufactured goods were exported or the distant destinations, both
urban and rural, to which Britain exported its surplus labour (e.g.
Pahl *et al.*, 1983; Robson, 1973; Lash and Urry, 1987: 94–9). Although
the contribution of colonial expansion to industrialization and capital
formation is acknowledged (Foster, 1974; Hobsbawm, 1969), the urban
and environmental implications of this are not followed through.

Yet just as the emerging industrial system of Britain assumed its
place in a developing international division of labour that is both social
and spatial (Massey, 1984; 1986) and one that equally comprehends
the agricultural and cash-crop production of the colonies, so also the
urban and environmental forms that result from this single, international system of production become component parts of a single and
global system of settlement and built environment. There is, in short,
a spatial hierarchy of production processes (Wallerstein, 1984) that
is expressed in the built environment.

The theoretical assumptions behind the analysis that follows,
therefore, are these:

1. Any system of socioeconomic organization (or mode of production) has a social division of labour.

2. This social division of labour is spatially and, generally, physically expressed in terms of building and, ultimately, urban form (King, 1984; 1986a).

3. Historically, a spatial division of labour, which in the early stages is expressed locally, later comes to be expressed regionally or nationally and subsequently, is expressed at an international scale (Feagin and Smith, 1987; Kentor, 1985).

For illustrative purposes, the following example is deliberately oversimplified.

Spinning and weaving in the peasant, pre-industrial economy of England is undertaken domestically by a household of, say, a woman and her husband and children in one room of a two-roomed dwelling that becomes adopted to this use: a necessary spatial division of labour results from the spatial requirements and location of the means of production (spinning and weaving equipment). Dietary requirements are supplied partly from food cultivated on the plot of land adjacent to the dwelling and cultivated by their own labour, and partly from the market.

The development of factory production under capitalism leads, in the earliest stage, not only to the classical spatial division between work and residence with the development of functionally specialized, industrialized towns, but (to keep the illustration simple) also to the development of specialized building forms — factories for the production process and particular — and different — forms of accommodation for labour. At first, these may be the densely packed courtyard dwellings of early industrialism, or row houses, dependent on region, and subsequently, the back-to-back dwellings or by-law terrace housing of the later industrial city.

Subsequently, diversification of the production process and markets leads to a spatial division of labour, which is expressed at a regional level. This implies functionally different urban regions and towns (Massey, 1979), each with functionally different building forms developing (King, 1984: 70-1): mills, factories, foundaries, and extensive working-class housing in production-oriented industrial towns; market halls, banks, offices, and more socially differentiated housing in more commercially oriented settlements; and theatres, assembly rooms, promenades, piers, and a variety of dwellings in the more spacious, consumption-oriented spas and resorts. Likewise, the emerging social structure of the regional and national society is manifest in a particular expression of residential building forms.

With the maturing of the international division of labour in the eighteenth and nineteenth centuries, the development of capital-intensive industrial production at the core (England) and labour-intensive agricultural production at the periphery (say India, South America, Africa), there is a resultant expression, not simply in the proportion of population urbanized in various parts of this division of labour but also, in the type of built environment that results. This means the dense concentration of living space (in back-to-backs, row houses, terraces) situated close to the factory at the point of production, in a heavily urbanized area, with no provision for self-provisioning in terms of food: the land or garden on which was grown the means of sustenance of the early-twentieth-century Lancashire or Yorkshire textile operative were not, as with their pre-industrial ancestors, located by their cottage or obtained from a regionally based market: as their staple diet now consisted of tea, sugar, cocoa, and wheat ('voluntarily' imported from the colonies), at least part of their 'garden' was located in India, Ceylon, the West Indies, West Africa, and Canada. Likewise, the cotton cloth worn by tea-plantation workers in Assam or Ceylon ('involuntarily' imported from Britain) was manufactured in the mills of Lancashire, and the machine tools and railway carriages used in Egypt or Natal, constructed in the workshops of Birmingham.

In short, an adequate understanding and explanation of the built forms and urban structure of any mode of production requires the simultaneous consideration of *all* elements of the social division of labour, irrespective of their geographical location. And this task is made easier when data on building form, function, and style are incorporated with information on the economic, social, and urban structure of particular places: social divisions of labour are architecturally expressed. The mansion of the West Indian plantation owner is a modified reproduction of his country house in Britain. Banking houses round the Empire initially express the style of the time they were built (Fermor-Hesketh, 1985). British urban development must be studied as part of, complementary to, and simultaneously with the *total* system of production and consumption in those parts of the world with which its economy, society, and polity were principally connected. One part is only comprehensible by reference to the other.

This analysis is informed by three related sets of theoretical literature: the study of urban systems, regional (and city) analysis, and the international division of labour and world-city systems. Pred defines an urban system as:

> a set of urban units (historically, cities: today, presumably metropolitan areas) that are interdependent, or bound together by economic interactions, in such a way that any significant change in the economic activities, occupational structure, total income, or population of one member unit will directly or indirectly bring about some modification in the economic activities, occupational structure, total income or population of one or more of the other members of the set.
>
> (Pred, 1980: 2)

This approach is demonstrated by Browning and Roberts (1980), whose work provides the best illustration of the first stage in the argument set out previously. In their study of the relation of British to Latin American urban development in the nineteenth century, they show how, within the emerging international division of labour, the distinctive nature of the occupational structure (the proportion of each country's working population in agriculture, manufacturing, and services) of both places can only be properly understood if one is seen as complementary to the other: the high degree of British industrial urbanization is the 'reverse side' of the commercial orientation of Latin American cities. And this analysis could no doubt be further extended to understand the architecture and built form of both places at this time (see King, 1990).

Browning and Roberts (1980) focus on economic and occupational structure in their analysis. In a conceptually related study, Salinas (1983) shows how spatial organization (the urban-settlement system, urbanization, and urban structure) in Peru is related over five centuries to different modes of production: mercantilism, industrial capitalism, and monopoly capitalism. Spatial organization in Peru is a direct outcome of the needs of mercantile Spain. The growth of commercial cities on the coast was directly related to the export of the agricultural surplus. As they grew richer, they attracted more of the rural elite 'to the residential comforts and glitter of urban life' (Salinas, 1983: 87). The change from feudalism to industrial capitalism, and the concomitant change from Spanish to English domination, brought no major alteration to Peru's internal productive structure or social relations: Peru continued its colonial mercantilist functions of channelling agricultural and mineral surplus to the metropolis (now England) though, with Britain's industrial revolution, Peru's exports were diversified, thereby integrating Peru more deeply into the European economy and making the new nation more dependent on the vicissitudes of European demand and business cycles (p. 90). English investment, however, also brought English cultural practices to the coastal towns: with the railway (supervised by British engineers) came the notion of the coastal resort (Schneider, 1986). In other Latin American states, such as Chile, similar processes also brought the cultural influences of British urbanism, including bungalow forms of the 1920s (personal communication, Mrs Christiane Collins, January 1986).

These latter illustrations draw on recent studies in regional political economy that link changes in the spatial organization of production (in relation to regions) to changes in the international division of labour: the reorganization of production on an integrated global scale dictates new roles for national and subnational economies (Salinas and Moulaert, 1983: 3; see also Feagin and Smith, 1987; Walton, 1985).

This view of British urban development as being part of (and dependent on) a larger systemic whole has much in common with Chase-Dunn's conception of the world-city system:

> Imagine a map of the earth in 1900. The outer boundary problem (of the system) is simplified at this point in time because nearly all areas of the earth have been incorporated into the system and we do not yet have to confront the sticky problem of the socialist states (which even today, despite their internal socialist structure, still remain within the capitalist world economy). For the moment, erase the national political boundaries that are drawn on the map and observe the cities, concentrations of human population in space. Now draw lines that indicate the commodity exchanges among the cities and towns of the world-system.
>
> What can now be observed is an exchange network among cities that has differential densities within it indicating various national and regional sub-systems, but that also exhibits a transitional structure similar in appearance to an airline route map. Exchanges among the largest cities of the core are dense both within and across national boundaries, while peripheral cities exchange mostly with core cities and very little with one another.
>
> (Chase-Dunn, 1985: 271)

Two further comments can be added to this. First, the date of 1900 is particularly valid for our analysis, not only because of the reason suggested by Chase-Dunn (1985) but also because this period (1900–1910) represents the height of British hegemony in the emergence of the capitalist world-economy from which time it has subsequently declined (Gamble, 1985); consequently, within the old international division of labour, the degree of industrial urbanization in Britain compared to that achieved by other competing economies at that time was greatest: moreover, much of the built environment, not least of London, had already been completed by this time and finally, the economic, occupational, and social composition of the metropolis was already largely determined by the city's location and function within this international division of labour (Ingham, 1984; King, 1989c).

Second, Chase-Dunn and other scholars focus particularly on the demography and political economy of urbanization in the world-economy. Yet there is equally a social, cultural, and ideational system within which such economic phenomena as 'labour markets' and 'price mechanisms' operate. A significant feature of this system is the physical and spatial form of the built environment, with its multiple meanings and properties, which forms the context within which economies and societies are transformed and gradually incorporated into the world-economy.

We have therefore, suggested three different fields that provide the prerequisite frameworks for understanding architecture and building form on a global scale: the study, in relation to the world-economy and an old and new international division of labour, of urbanization (Feagin and Smith, 1987; Glickman, 1987; Kentor, 1985; Timberlake, 1985; 1987), regional development (Salinas and Moulaert, 1983), and cities (Chase-Dunn, 1985; Pred, 1980).

BUILDINGS AND THE INTERNATIONAL DIVISION OF LABOUR

This framework can now be applied to an understanding of some of the changes taking place in contemporary Britain. Clearly, this is meant as a broad-brush approach, details of which need to be carefully worked out.

In the old international division of labour, Britain became the most urbanized country in the world, with 80 per cent of its population in towns at the height of the imperial connection (1914). As suggested in Chapter 1, this development presupposed the existence of particular raw materials (especially cotton and wool, but also rubber, tin, timber, minerals, sugar, and others), which were processed in urban (especially port) centres, as well as food imports (wheat, rice, etc.) and assured markets for manufactured products. As Christopher (1988) and Massey (1986) point out, specific sectors of the economy and specific regions and towns exhibited the phenomenon of 'dependent urbanism' more than others. (In one sense, the most dependent city was London, which largely monopolized many of the financial and banking functions of the world.) While the proportion of British trade, with its overseas Dominions and Empire, varied over the period, in the 1930s, two-thirds in value of British exports were destined for dominion and colonial markets and, as discussed, significant proportions of the exports of individual colonies came to Britain. In addition, something under half of overseas direct investment (in the region of £4,000 million) was placed in the Empire (Christopher, 1988; Dunning, 1983). British urbanization, therefore, was a specialized and symbiotic part of a colonial space economy as well as the world-economy in general.

This meant, therefore, not only a particular economic and occupational structure, but also, the existence of a vast building stock, supported and maintained by the profits of industrial production derived from her pre-eminent position in the world-economy. Whilst much of this stock was in the form of urban working- and middle-class housing, a significant proportion of this globally (as well as nationally) derived surplus had been creamed off over the previous century and invested in substantial aristocratic and bourgeois dwellings, as well as other building forms — in the metropolis, on the edge of cities, and especially in 'the country'. Without Britain's privileged place in the world-economy, and without its imperial advantage, it is unlikely that such a large stock of such building would have developed. The archetypical example is the large country house of an industrial baron deriving his profits from imported Peruvian guano (see Girouard, 1971).[2]

As Britain's place in the world-economy declined during the middle decades of the twentieth century, a large part of this elite stock of mansions and country houses disappeared, there being neither the capital nor an 'appropriate' distribution of economic and social power to sustain them. In 1974, Cornforth estimated there were between 1,500 and 2,000 'notable' country houses in Britain. Between 1945 and 1974 some 340 of such 'notable' houses were demolished (Cornforth, 1974: 4). As further deindustrialization occurred from the 1960s, Britain became increasingly 'over-urbanized' (the term is used with its original meaning of the 1950s, i.e. compared to the historical development of urban industrialized countries in the West, many developing countries were said to have a higher degree of urbanization than their level of industrialization (measured by the proportion of the workforce in the industrial sector) could support. See Sovani, 1966).

There is, in short, no industrial economy operating (as it did in the nineteenth and early twentieth centuries when much of this built environment was constructed) in a globally privileged position to keep that environment maintained. This applies especially to those working-class towns of the industrial Midlands and North as well as the consumer seaside resorts that depended on them (Urry, 1987). But whilst many of these older, industrially based dwellings and buildings go out of use, the shift to the services economy privileges parts of that old environment (typically, the interwar suburbs of detached and semi-detached housing) and allows them to be retained.

However, if the UK is overurbanized in terms of its industrial housing stock at the national level, in the new international division of labour of the 1970s, there is insufficient urban housing for the growing numbers of internationally migrant labour. Hence, migrants ('peasant' and urban) from the post-colonial periphery (in this case, the Caribbean, Bangladesh, India, Pakistan) improve their economic and housing conditions by moving to old working-class housing at the core. More economically signficant, however, is the role played by the elite environment of mansions and country houses in the new international division of labour of the 1980s.

Here, because of the country's earlier privileged international position and cultural practices governing preferred life styles and capital investment among the elite, Britain has an enormous and probably unrivalled stock of country (and urban) properties unaffected — as was much of Europe — by wartime destruction, many of them invested with land, history, and a negotiable social meaning. As, in many cases, their original construction drew on internationally derived capital, so too does their current refurbishment and use. Britain's services role in the new international economy (tourism, as well as financial, corporate, and producer services) injects new money into an old environment: the residential property of an earlier elite becomes a

137

source for investment and status, whether for world-city financiers, oil-rich Middle Eastern investors, or American companies looking for platforms to Europe (see King, 1990; also *Saville's Magazine,* 1985; 1987; *Property International,* 1985; Thrift, 1987b). For such properties in the 1980s, local markets have become national and national ones, international.

COLONIAL URBANIZATION AND BRITAIN

Some of the gaps in the historical understanding of urbanization in Britain have been filled in recent years by de Vries (1983) who, in a study of European urbanization between 1500 and 1800, makes a number of revisions to conventional assumptions. He demonstrates, for example, that a complex international urban system in Europe pre-dated the era of industrialization and that this sytem was largely consolidated in the seventeenth century; in brief, the construction of an urban system was a precondition, not a result of modern industrial growth. The evidence is provided by a study of changes in the urban population in 379 West European cities with populations over 10,000 in the period studied.

De Vries makes the point that the city's function in economic growth can be understood only in a regional context and, whilst recognizing that the nation-state does not provide the relevant unit of analysis, is not as forthcoming as he might be on what exactly are the boundaries of a particular space economy. The broad generalization, though qualified, is that Europe, though never a single polity, nevertheless had 'a real identity and a real unity. That unity was religious and cultural, but also economic' (de Vries, 1983: 84). After showing how the urban core of Europe moved from the Mediterranean, via the southern to the northern Netherlands between the sixteenth and eighteenth centuries, he writes:

> The urban civilization that emerged in Flanders supported itself as head of a northern commercial system; that of the Dutch Republic built upon this, presiding over a far-reaching European and *colonial system* that succeeded in subordinating all rivals; that of Britain took this over and added to the critical dynamic of the Atlantic economy. By 1800, Europe's urban core displayed a more pronounced maritime orientation than ever before, the culmination of a centuries long process.
>
> (de Vries, 1983: italics added)

The question is, however, whether by 1800 the European urban system that de Vries charts can be treated more or less as an autonomous unit, not least in terms of Pred's (1980) concept of an 'urban system', which de Vries uses to inform his analysis. To take the case of Britain in particular. Between 1650 and 1750, four out of five of Britain's largest cities were port cities (London, Edinburgh, Bristol, Newcastle). In the later eighteenth and early nineteenth centuries, the simultaneous and complementary growth of Britain's ten largest industrial and port cities (the majority in the north) not only suggests a powerful interdependence between production and import/export but also, their collective interdependence with another set of port cities and inland towns overseas (see Table 7.1).

Table 7.1 Ten largest cities in Britain, 1650–1800 (,000)

1650		1700		1750		1800	
London	400	London	575	London	675	London	865
Edinburgh	35	Edinburgh	40	Edinburgh	57	Edinburgh	82
Bristol	20	Norwich*	29	Bristol	45	Liverpool	78
Norwich*	20	Bristol	25	Norwich*	36	Glasgow	77
Newcastle	13	Newcastle	14	Newcastle	29	Manchester	70
York*	12	Exeter	14	Glasgow	24	Birmingham*	69
Exeter	10	Glasgow	13	Birmingham*	24	Bristol	64
Yarmouth	10	York*	11	Liverpool	22	Leeds*	53
Oxford*	9	Yarmouth	10	Manchester	18	Sheffield*	46
Worcester*	8	Colchester	10	Exeter	16	Plymouth	43
		Sheffield*	10	Leeds*	16		
		Chester*	10				
		Shrewsbury*	10				

Source: de Vries, 1983: 270–1.
Note: * Inland cities

The critical role of these ten cities in the emergence of the old international division of labour, based on the exchange of raw materials from underdeveloped countries for British manufactured goods, is sufficiently manifest by their economic decline from the 1930s and the collapse of much of their industry from the 1970s. If it was not previously acknowledged that the rapid growth of these cities in the eighteenth and nineteenth centuries was dependent on Britain's privileged role in the international economy (*vide* Robson, 1973) their subsequent decline has provided the evidence.

By contrast, some of the largest towns of the mid-sixteenth century (Norwich, York, Exeter, Worcester, and Coventry), the products of an earlier mode of production and part of a local, regional, and mainly national (and European) spatial division of labour (including relative self-sufficiency in food supplies) are now flourishing. Having been bypassed by the ravages of colonially related industrialization, these historic centres are now bases for the new service-oriented economy.

It is, however, London and the large port cities of 1800 (the 'colonial port cities' of Britain)[1] that demand our attention. Given the massive growth of London, the port cities and industrial cities that both supplied and were supplied through them (Manchester, Birmingham, Leeds, and Sheffield) in the late eighteenth and nineteenth centuries, it is clear that, by 1800, a whole new set of other port and inland cities was equally, if not more part of an emerging British colonial urban system than were particular cities in Europe: they would include Calcutta, Bombay, Madras, Dacca, Nassau, Kingston, Sydney, Halifax, Montreal, Toronto, Port of Spain, Bridgetown, Gibraltar, and others.

A century later (1900) this colonial urban system had expanded to include Aden, Hong Kong, Cape Town, East London, Durban, Pretoria, Johannesburg, Salisbury, Blantyre, Mombasa, Salisbury, Kampala, Zanzibar, Lagos, Accra, Nikosia, Suez, Port Louis (Mauritius), Mahé, Kuching, Georgetown (Guiana), Melbourne, Brisbane, Adelaide, Perth, Hobart, Christchurch, Wellington, Port Moresby, and Port Stanley (Gill, 1901; see Table 7.2). What is significant about these cities is that their built and spatial environments (as well as other phenomena) begin to have more in common with each other than each has with the economically, politically, and culturally very different environments of the interior of the countries and continents where such ports were located.

Moreover, if the purpose of theoretical concepts (such as 'urban systems') is to help explain contemporary realities, whether these are the social and ethnic composition of London, the nature of its East End sweatshops (King, 1990), the predominance of British free-market physical-planning practices round the world, or even the complementary similarities of different physical and spatial environments (see Chapters 1 and 2) it is clear that a notion of 'urban system' must take account of geopolitical history. As much of the present structure of Britain's (and London's) urban development was already in place by 1900, it is worth examining Britain's place in a larger imperial division of labour at this time, using both Pred's (1980) and Chase-Dunn's (1985) conceptualizations and looking at the colonial urban system of which London was the apex and in which other major British cities (for

Table 7.2 Major cities in the British colonial urban system, 1800 and 1900

Colony	Date of incorporation		Major city		Population in 1900 (,000)
	pre 1800	*pre 1900*	*pre 1800*	*pre 1900*	
EUROPE					
Gibraltar	1704		Gibraltar		25
Malta		1800	Valetta		60
Cyprus		1878	Nikosia		13
ASIA					
Aden		1838	Aden		42
India	1612–		Calcutta		1027
			Bombay		776
			Madras		509
				Delhi	209
Ceylon	1796		Colombo		127
Straits			Penang		
Settlements	1785			Singapore	512
		1819		Malacca	Δ
		1843		Hong Kong	254
Sarawak		1888		Kuching	Δ
Labuan		1846		Victoria	Δ
North Borneo/					
Brunei		1881			
Brunei		1884			
AFRICA					
Cape Colony		1815		Cape Town	79
				Port Elizabeth	23
				Durban	27
Natal		1856		Pietermaritzburg	18
Transvaal		1900		Pretoria	12
				East London	7
Basutoland		1868			
Bechuanaland		1868		Vryburg	Δ
Rhodesia		1888		Salisbury	Δ
Central Africa/				Blantyre	Δ
Nyasaland		1889		Zomba	Δ
Gambia	1664			Bathurst	14
Sierra Leone	1787			Freetown	Δ
Gold Coast/ (Ghana)		1868		Accra	Δ
Lagos		1861		Lagos	33
Nigeria		1886			
East Africa/ with Uganda		1888		Mombasa	Δ
Zanzibar		1888		Zanzibar	30
Somaliland		1884		Berbera	Δ

Colony	Date of incorporation		Major city		Population in 1900 (,000)
	pre 1800	pre 1900	pre 1800	pre 1900	
St Helena	1673		St Helena/ (Jamestown)		4
Ascension		1815	Georgetown		Δ
Mauritius		1810	Port Louis		65
NORTH AMERICA					
Canada	1623–		Montreal		268
			Toronto		208
			Quebec		69
			Ottawa		60
			Halifax		41
Newfoundland	1583		St John's		31
SOUTH AMERICA					
British Guiana		1803	Georgetown		49
Falkland Islands		1833	Port Stanley		†
WEST INDIES					
Bermuda	1609		Hamilton		2
Bahamas	1670		Nassau		Δ
Jamaica	1629		Kingston		Δ
Leeward Islands	1626–		St John (Antigua)		Δ
Windward Islands	1605–		St George's/ Grenada		Δ
Barbados	1605		Bridgetown		Δ
Trinidad & Tobago	1797		Port of Spain		34
Honduras	1783		Belize		7
AUSTRALASIA					
New South Wales	1787		Sydney		99
Victoria	1787		Melbourne		478
Western Australia		1829	Perth		38
South Australia		1836	Adelaide		162
Queensland		1859	Brisbane		121
Tasmania		1803	Hobart		Δ
New Zealand		1841	Wellington		47
			Auckland		67
Fiji		1874	Suva		Δ
British New Guinea		1884	Port Moresby		Δ
Cook Archipelago		1888	Raratonga		Δ

Source: Extracted from Gill, n.d. *c.* 1901.

Note Δ Not available.

　　– Hyphens indicate '1612 onwards'

　　† Total island population = 1,790.

example, Liverpool, Glasgow, and Manchester) played a major, but essentially dependent role.

Of course, this colonial urban system was neither autonomous nor exclusive but was embedded in a larger world-economy: the colonial economy accounted for only a portion of Liverpool's or Glasgow's economic base (though a large one) and London had existed prior to the rise of the Empire just as had other cities in the Far East. It is, however, the *extent* to which any city was created by, or became dependent upon, the colonial political economy that needs to be examined. In some cases this dependence was great.

THE BRITISH COLONIAL URBAN SYSTEM: A TOPOGRAPHICAL DESCRIPTION

By the beginning of the twentieth century, this colonial urban system was extensive. It linked the interior of countries to their ports and the ports both to each other and the metropolis. It was the system by which many countries were brought into the capitalist world-economy. It provided the nodal transportation links for the import and export of goods and services and was the network for distinctively colonial forms of international labour migration (including indentured labour). It established labour markets, but also, with the transformation of old and the creation of new environments, it provided the physical and spatial infrastructure for the restructuring of the social, cultural, and political order, creating centres for new modes of consumption, and the transformation, through 'modernization' and commodification, of social, cultural, and political consciousness. Table 7.2 (p. 141) lists the major towns and cities of this system, the date at which the territories of which they were a part came under British political and cultural control and their approximate population size in 1900. A glance at the cities and countries in this system indicates both the historic and present importance of these cities in the development of both Britain's economy, its subsequent ethnic composition and culture, as well as the larger world system. Given their importance, therefore, some historical notes from Gill's *The British Colonies, Dependencies and Protectorates*, published in about 1901, are apposite.

In the early twentieth century, the population of India (including independent and feudatory states) was 297 million, of whom 121,000 were European, principally British, and largely resident in the main cities or other urban areas. As Palat *et al.* (1986) point out, as India

143

was incorporated into the world-economy, it became lacked into a different international division of labour, importing manufacturing goods (principally from Britain) and exporting raw materials, to the benefit of British interests. The expansion of commercial crops, largely produced for the world market, coincided with a long-term stagnation in foodgrain production (pp. 184–5). Gill (c. 1901) indicates that between 1850 and 1900, the value of British trade with India grew by a factor of seven, from £30 million to £213 million; one-third of all India's exports were taken by the UK but three-quarters of India's imports came from there, mainly cotton goods, bullion, machinery, metals, railway carriages, ships, boats, and woollens. In return, India sent raw cotton, cotton goods, jute, rice, hides and skins, tea, wheat, and coffee. As the great jute-producing country of the world (mostly cultivated in lower Bengal round Dacca and Calcutta) India sent most of its exports to be processed in Dundee.

Loans raised on the London market were either to the government of India, for the Indian railways, or to the British agency houses that owned or managed a large number of plantations, factories, transport services, or banks. Palat *et al.* point out that in 1911 Europeans (mainly British) owned some 88 per cent of tea plantations, 93 per cent of indigo plantations, and 35 per cent of collieries. European agencies managed 58 per cent of Indian-owned industrial enterprises (Palat *et al.*, 1986: 186).

These agencies, as well as the whole apparatus of colonial management and economic and political control were located in the cities, principally, Calcutta which, according to Gill (*c.* 1901), was 'adorned with many stately edifaces' and 'so unites the luxury of East and West that it has often been styled the "London and Paris of Asia" '.

Bombay, the third-largest city of the Empire, was 'the busiest port in Asia', responsible for shipping three-quarters of India's cotton; by 1901, there were some 190 cotton mills in India, two-thirds of them in Bombay (the first had been established in 1854). These mills concentrated on yarn production for export and for domestic handloom weavers, complementing British interests, which monopolized the more profitable, finer count piece goods and still expanded production (Palat *et al.*, 1986: 186).

By the mid-nineteenth century, Bombay had also become the financial centre for British India and in the next decades, its urban structure and port facilities were remodelled to orient its economic function more closely to the needs of British colonial interests

(Dossal, 1989). The change was also to be seen in its architecture, where new colonial institutions (a railway station, a town hall, banks, and buildings of commerce) replicated the forms and styles of those in the metropolis (Darley, 1985) as well as expressing imperial power (Metcalfe, 1989).

The third largest, city, Madras, exported largely coffee, sugar, indigo, dye, and cotton. Here, too, urban remodelling had taken place with a pier, a new harbour, and railway facilities constructed at the end of the century by the British administration.

This economic and trading interdependence was manifest, therefore, in the urban and especially port structure of *both* countries, which were linked by the main steamship routes, in the metropole, running from London (Woolwich) and Liverpool.

In Ceylon (Sri Lanka), over half of the trade was with the UK which took nine-tenths of her tea, and two-thirds of her coffee, trade that largely accounted for the rapid growth of Colombo (127,000 population in 1900). In 1900, Ceylon had most of its agriculture (coconut, paddy, tea, and coffee) in plantations largely serviced by 'Tamil coolies' brought in by the colonial administration from South India and who formed 25 per cent of the population.

Within the imperial system, the Straits Settlements (Singapore, Penang, and Malacca) performed a key economic and political role. Singapore's main function was strategic, commanding commercial channels to the East Indies, China, and Japan and therefore hosting the headquarters of British military and native forces. It was also, however, a vast commercial city, the entrepôt for the produce of the surrounding countries (the Malay Peninsula, the Dutch East Indies, Japan, Borneo, Siam, and the Philippines). About one-sixth of all the trade of the colony was with the UK.

Like Singapore, Hong Kong's two major imperial functions were as military and naval station (about 3000 troops were based there in 1900) and as a huge commercial entrepôt. At that date, the population was 254,000, 90 per cent of whom were Chinese. The port of Victoria (Hong Kong) did more trade than any other in Asia, controlling commerce with China and Japan. However:

> Hong Kong is so small it has nothing of its own production to export: its merchants, mostly Chinese and British, are engaged in collecting the surplus products of surrounding districts and distributing them in exchange for European and other wares.
>
> (Gill, *c.* 1901: 115)

In 1899, British trade with Hong Kong was worth £3½ million, exporting cotton, woollen, and iron goods, copper and lead, and importing silk, tea, hemp, and copper. The total export trade of Hong Kong was estimated at £25 million: and again demonstrating the interdependence of the ports mentioned, Gill (c. 1901; the source of these data and quotes) compares the tonnage entered and cleared at the port of Victoria in 1898 at over 17 million tons (over half of which was British) with the 29 million tons in London and 19 million tons at Liverpool.

What is now South Africa included, in 1901, the separate colonies of the Cape, Natal, and Transvaal. The population of the Cape Colony was about 2.5 million, 'of whom only 377,000 are white'. The Colony was 'especially rich in diamonds and copper but the mining industry was still in its infancy'. However, annual turnover of trade in 1899 amounted to almost £43 million, exporting mainly bullion, diamonds, and wool in return for textile fabrics, machinery, hardware, agricultural implements, foodstuffs, furniture, etc.

Natal, incorporated into the empire in 1900, exported gold, wool, and hides, with three-fifths of its entire trade done with the UK and most of the foreign commerce undertaken through Durban.

The Transvaal 'the land of gold', had 'attracted miners from all over the world'. Of the 900,000 population 'one third are white'. In less than fourteen years, Johannesburg (population 102,000) had risen 'from a few scattered huts made from paraffin tins to its present position as one of the largest gold mining centres of the world. It has some remarkable fine streets and the buildings, shops and stores compare favourably with many old towns'. In the capital, Pretoria (12,000 population), the streets were 'arranged in parallelograms all bordered by magnificent willow trees, originally planted as fencing poles' (Gill, c. 1901). In Rhodesia (Zimbabwe), with the capital at Salisbury (Harare), the British South African Company had invested in gold production, as well as silver, copper, and land for cereal production. The principal exports from Mombasa, British East Africa (Kenya) were tropical products — bananas, arrowroot, casa, coffee, india rubber, etc. — and these were exchanged for Lancashire and Bombay cotton cloths. From Freetown, Sierra Leone, kola nuts, ground nuts, india rubber, and hides were sent to Liverpool in exchange for 'Manchester goods'.

Likewise, from Accra on the Gold Coast (Ghana), a Crown Colony of about 1.5 million population (including 150 Europeans), were sent india rubber, palm oil, gold dust, timber, three-quarters of the trade being with the UK. Again, it is worth pointing out that 'the commerce of the West African coast is chiefly in the hands of Liverpool merchants' (p. 204).

In 1901, Northern and Southern Nigeria were separate provinces from Lagos. 'The stream of British commerce with this vast region flows towards London and Liverpool, the headquarters of the merchants who form the Royal Niger Company'. The Company had established over 100 factories on the main river as far as Egga, with other trading stations established at Yola, Lokoja ('the headquarters of several Liverpool merchants') and Loko. The chief exports were rubber, ivory, palm oil, kernels, gum arabic, etc.; the main imports, textile fabrics, hardware, earthenware, tobacco, guns and powder.

Though the majority of Canada's trade was with the USA, in 1901, 40 per cent of it was still with the UK. Britain mainly imported timber, grain, cheese, horses, and other live animals, meat, etc., and exported to Canada woollens, metals, cottons, apparel, silks, spirits, books, stationery, railroad bars and engines, etc. The major ports of Montreal (population, 268,000), Toronto (208,000), Halifax (41,000), and Quebec (69,000) were supplemented by Vancouver (26,000), linked to Hong Kong by the Canadian Pacific steamboat.

Britain's oldest colonies were in the Caribbean and, although economically underdeveloped in the early twentieth century, and with the largest part of their trade with the USA, their importance was in providing British shipping with a lucrative business and again, establishing the link with Liverpool. Jamaica, first acquired in 1629, had trade worth some £4 million with the UK, exporting fruit, coffee, sugar, rum, and importing cotton goods; half of Dominica's trade was with the UK; one-third of that of the Windward, Trinidad and Tobago Islands, exporting, through the Port of Spain (34,000 population), raw sugar, cocoa, and asphalt; a quarter of Barbados trade was with the UK, one-half of that from British Honduras (exporting logwood and mahogany).

The value of exports from the Falkland Islands was low (at £200,000) and consisted of sheep, wool, and frozen mutton.

The final major unit in this imperial division of labour (not all parts of which have been mentioned) was Australia which, in 1901 had a population of 4.5 million, 'mainly British', with 200,000 aborigines and 45,000 Chinese immigrants. The trade of New South Wales was 'larger than that of any British possession excepting India and Canada, of an annual value of £54 million, six-sevenths of which was with various sections of the British Empire' (Gill, *c*. 1901). Half of its wool exports went to the UK as well as one-tenth of its gold. The commerce of the UK with NSW in 1900 was worth £19.5 million, with Britain exporting metals and cottons, woollen goods, etc. Likewise, one-third of Victoria's trade was with Britain: 'The colonists have established intimate trading relations with London and Liverpool'. Queensland, nearer to China, India, and California, exported only half of its annual exchange of £18.5 million to Britain but imported telegraphic wire, nails, metal goods, etc. Western Australia's £15 million of trade was 'growing rapidly', most of it conducted with the UK, again exporting wool and importing machinery, with Perth a city of 38,000. Three-quarters of New Zealand's trade (worth about £21 million) was with the UK.

The justification for this lengthy (yet also curtailed) account of Britain's trading relations with her imperial possessions is, to reiterate the point, to demonstrate how far British industrial urbanization — its degree, location and distinctive built environment — was, within the larger world-economy, strongly influenced by and, in places, dependent upon a colonial system of production: it was largely produced by it and cannot be understood except as part of it. (Nor has this description taken account of the 'informal' colonial input from the Argentine and other Latin American states in the same period.) Admittedly, much of the argument made here is by inference rather than proof. But if concrete evidence is needed to demonstrate the complementary and interdependent development of this urban system it can be found, literally, on the ground, in the urban infrastructure of the building, architecture, and urban forms of the imperial and colonial system (Darley, 1985; Fermor-Hesketh, 1985; King, 1976; Morris, 1985), irrespective of more conventional economic and social histories. It is this that gives these colonial cities more in common with each other than any of them has with the 'traditional', pre-capitalist forms of built environment that exist in the country's indigenous interior. And the infrastructure of these cities formed the bases for a later transformation to the present world-economy.

Moreover, the colonial system was not just an economic but also a political, ideological, social, and cultural system, dimensions that need to be recognized if we wish to understand the way in which strands of yesterday's colonialism are woven into, and influence the fabric of today's world-economy and politico-cultural system. And whilst the colonial urban system was not 'sealed', nor did it operate in territories that had previously been a *tabula rasa*, it none the less continues to exist as a set of powerful influences, not least in terms of language, institutions, and practices that influence the contemporary world and its international system of cities (King, 1989c).

In historical studies of the planning and architecture of Canadian, American, South African, or Australian cities, it is conventional wisdom for scholars to look at contemporary developments in Britain as a major (though not the only) source of understanding early mercantile and industrial capitalist forms of development (see, for example, R. Irving's (1985) 'Georgian Britain' in *The History and Design of the Australian House*; Mullins, 1981; Sandercock, 1975). Such analyses, however, go only half-way; colonial urban planning or architectural forms (or, for that matter, legal, constitutional, literary, or social and cultural forms) are not simply *derivations* of core forms but rather, functionally interdependent parts of a single system. Such approaches are none the less in advance of much urban historical analysis in the UK where processes of industrial urbanization and the built environment it produced are frequently treated as developing quite independently of other parts of the imperial and world space economy.

CONCLUSION

The assumptions behind this, as well as earlier chapters, are twofold: that urban and other phenomena can only be adequately understood by treating them as part of a larger world-system, of economy, society, and culture, of which they are an integral part. And second, that the built environment, in all its various conceptualizations, is both a product of and a major resource for understanding these global processes. As indicated in the preface, what these chapters have attempted is to develop theoretical tools for understanding these processes, which also transcend some of the disciplinary boundaries that divide, for example, sociology from geography, architecture from political economy, and inhibit understanding.

However, no theory develops in a vacuum; research needs to be grounded in data collection, informed, of course, by hypotheses and theory, the construction and reconstruction of which must be utilized to suggest new frameworks, theories, problems – and solutions.

NOTES

CHAPTER TWO

1. Cf. Gilbert A. Stelter: 'We must look in a comparative way beyond Canada to the cities of the US and other countries which were the products of European expansion if we wish to fully understand our own experience.' Stave (1980) 'A conversation with Gilbert A. Stelter: Urban history in Canada', *Journal of Urban History* 6: 177–209.
2. This listing of characteristics from Telkamp's essay (1978) is given in more detail than in the 1980 version of the paper (published 1985).
3. Johnston, *City and Society* (and other authors) divide such societies on the basis of socioeconomic organization as reciprocal exchange and rank-redistribution (Johnston, 1980: 38–9). For our purpose here, a more detailed categorization would be necessary.
4. Horvath (1969) uses the term 'intervening group' for populations imported from a third territory in the colonial system, such as, for example, the Asians in Uganda.
5. I am indebted to Deryck Holdsworth for this example; equally trivial, though useful as a symbol, is the culturally distinctive artefact of a 'tropical ukelele', sold by a Welshman's music store (branches in Calcutta and Mussoorie) found in the north England commercial-industrial city of Leeds.
6. It seems necessary to mention this only in the context of much economic determinism in studies of contemporary urbanism in the West. Significantly, the Middle East and Latin America are often omitted from these studies.
7. For a bibliography on these, see R.R. Reed (1976), *City of Pines. The origins of Baguio as a colonial hill station*, Center for South and South-East Asian Studies, University of California (Berkeley).

CHAPTER THREE

1. For example:

I moved the HQ from that close, unhealthy and altogether hateful
spot Kampala to a lovely place on the Lake; two great grassy
hills, like the Kingsclere Downs, rising almost straight out of the
water; and a view over the Lake like over the sea dotted with a
dozen islands. I put the European quarters on the highest hill, and
the Soudanese troops on the lower one, and we marked out all the
streets and divisions giving each man a small compound, and
established a market place, and cut great wide roads in every
direction. Before I left there was already quite a neat town of
about 1000 inhabitants, ten times more healthy than at Kampala
. . . The name of the settlement is Port Alice.

(Portal, Sir G. (1894) *The British Mission to Uganda*, (subsequently to
be Entebbe, Uganda). Quoted from *Colonial Building Notes*, no. 42,
January 1957, p. 13.)
Nairobi was established in 1896 as a railway capital, laid out by a
sergeant of the Royal Engineers (*Overseas Building Notes*, no. 141,
December 1971). The site was 'badly chosen . . . the same mistakes
of overcrowding and insanitary conditions as existed in England in
the early nineteenth century have a tendency to repeat themselves in
new countries' (*Report on Sanitary Matters in the East Africa Protectorate,
Uganda & Zanzibar*, 1913 quoted in White, L.W.T. *et al.* (1948)
Nairobi: Master Plan for a Colonial Capital.

2. National Independence dates in Africa: South Africa, 1909; Egypt,
1922; Libya, 1951; Tunisia, Morocco, and the Sudan, 1956; Ghana,
1957; Guinea, 1958; Mauritania, Mali, Senegal, Togo, Dahomey,
Ivory Coast, Upper Volta, Niger, Nigeria, Chad, Somali, Cameroun,
Central African Republic, Gabon, Congo-Brazzaville, Congo
Republic, and Malagasey, 1960; Sierra Leone and Tanzania, 1961;
Ruanda, Uganda, Burundi, and Algeria, 1962; Kenya, 1963; Zambia
and Malawi, 1964; Gambia and Rhodesia (UDI), 1965; Lesotho,
Botswana, and Swaziland, 1967; Ifni, 1969; Guinea-Bissau, 1974;
Mozambique, Cape Verde, the Comoros, São Tome and Principe,
and Angola, 1975.

3. This is an important distinction. The expatriate elite did not simply
'replicate in the capital the society of their own homeland' (Lloyd
(1979) *Slums of Hope. Shanty Towns of the Third World*, p. 19) as is
frequently stated. Had they done this, there would have been little
point in them going. The colonial 'third culture' was a subtly
modified version of the metropolitan (see King (1976) *Colonial Urban
Development*, pp. 58–66 and Chapter 6).

4. ' "Town and country planning" has been a borrowed phrase which
is still much in vogue in India although it appears to be going out of
use in the language of its origin in the United Kingdom' (Manzoor
Alam (1972) *Metropolitan Hyderabad and its Region. A Strategy for
Development*; see also Parkin (1972) *Town and Country in Central and
Eastern Africa*). On the basic conceptual distinction, its incorporation

into linguistic categories, and the extension of 'town-country' relations globally, see the seminal works of Williams (1973) *The Country and the City* and (1976) *Keywords. A Vocabulary of Culture and Society*.

5. The first of these, established at Edinburgh University, was followed by the setting up of other centres at Nottingham, UWIST, Sheffield, and Newcastle Universities. Other courses relating to aspects of regional and 'development' planning were to be established at Swansea, Birmingham, East Anglia, and Sussex Universities. In addition, countless other students from 'developing' countries attend courses in polytechnics and institutes of further education in subjects related to environmental management, such as housing administration and quantity surveying.

CHAPTER FOUR

1. Britain developed as an essential part of a global economy, and more particularly as the centre of that vast formal or informal 'empire' on which its fortunes have so largely rested. To write about this country without also saying something about the West Indies and India, about Argentina and Australia, is unreal.
   (Hobsbawm (1969) *Industry and Empire*, Penguin edition 1970: 20)

   That events having their origin in connection with the first three of these countries very considerably affected the political and hence, economic fortunes of the UK in the early 1980s seems hardly worth mentioning, except perhaps to give substance to the point made by Walton that studies of urbanism have often focused on the city in isolation from its larger national and international context, though 'much of what is often crucial in explaining local phenomena is extra-local in origin' (Walton (1976) 'Political economy of world urban systems: directions for comparative research', in Walton and Masotti (eds) *The City in Comparative Perspective*, pp. 301–13, 309).

2. In Fraser (1980) *A History of Modern Leeds*. The inclusion of a chapter on 'Imperialism and Leeds Politics' does not substantially detract from the viewpoint expressed here. Whilst considerable reference is made to other immigrant groups, namely, the Irish and Jews, there is little discussion of why or how they arrived there; as with other studies of urban immigrants, their presence is simply taken for granted. An additional irony is that the authority on the town's eighteenth-century economy left for the West Indies, and then went to Australia (p. 43) whilst the joint author for the chapter on nineteenth-century industrial development was, at the time of publication, at the University of Dar es Salaam.

   The criteria for choosing this book were principally, though not entirely, personal: as I previously lived in the city, I am interested in it.

3. A rough count of the research articles appearing from 1977 to late 1982 suggests that the majority (about fifty) reported work on Europe (excluding the UK), especially France, about half that number referred equally to work on the USA and the UK; about a dozen are on Latin America. Other than a fair number on general theory, fewer than four to five each seem to be on the socialist countries of Europe, or on Australasia, Africa, South or South-East Asia, or the Middle East. Subsequent work began to look at the political economy of global urbanization. These lacunae had been pointed out prior to the meeting of the Research Committee in 1982. There had been a 'degree of ethnocentrism' in the new urban sociology of the 1970s, an overconcern with the so-called urban crisis in the capitalist core countries; insufficient attention had been paid to the effects of exploitation of the Third World through multinational corporations. A new focus was recommended on methods of comparative and historical analysis: I. Szelenyi (1981) 'Structural changes of and alternatives to capitalist development in the contemporary urban and regional system', *Int. J. Urban and Regional Research*, 5 (1): 1–14, 2–3; papers from the meeting were published in subsequent issues of the *IJURR*. The perspective of the journal was to change in 1987, with a change in policy to include 'urban history, economic restructuring and urban development, Middle Eastern and African urbanization and the relationship between structure, consciousness and action in urban social theory'. See 'Editorial', *International Journal of Urban and Regional Research*, 11 (1), p. 2, 1987.

CHAPTER SEVEN

1. Much of the writing on the phenomenon of colonial urbanization suggests that it is something which only applies to the analysis of urban phenomena in Latin America, Asia or Africa. Even to incorporate Canada and the United States, Australia or New Zealand into the same framework is not always easy to accept, but to include France, Spain, the Netherlands, Belgium or Britain is seen as even more problematic.

    Yet what is the difference, conceptually, between, on one hand, the rise of 'colonial port cities' (Basu, 1985), the primate cities of the Far East (Murphy, 1969), or the restructuring of North African urbanism from the ancient inland market centres of Fez and Meknes to Casablanca and Rabat as a result of French colonialism (Abu-Lughod, 1976; 1980) or, in West Africa, from inland trading and industrial centres such as Kano to Port Harcourt (Mabogunje, 1980) or, in India, from ancient centres of manufacturing, like Ahmedabad (Naqvi, 1968) to Calcutta, Bombay or Madras, from, on the other hand, the shift in Britain of population from urban centres such as Norwich, York or

Salisbury to Liverpool, Glasgow and London? It is all part of the same process, except that as different academic specialists treat the phenomena separately, and give them different names, we are actually prevented from seeing what is taking place. In Britain, it is called 'the development of overseas trade'; in Morocco or Malaya, 'colonial development and exploitation'.

(King, 1986a)

Of course, the critical difference between industrial production at the core and the supply of raw materials and markets on the periphery would need to be addressed.

The extent to which Britain's industrial revolution depended on overseas trade, especially in the nineteenth century and with regard to particular industries is discussed in Thomson (1982).

2. An equally good example is provided by the domestic property assets of international financier Sir Ernest Cassel whose extensive interests ranged from Indian steelworks, Mexican railroads, Swedish iron mining, South African mines to a variety of investment activities in Latin America, Turkey, China, Japan, Egypt, Morocco, and other places. It was these that helped finance his various English houses, a stable at Newmarket and the costly renovation of his marble-lined Park Lane mansion (Thane, 1986).

The country house was often the favoured destination for this overseas-derived wealth. Such country houses (as opposed simply to 'houses in the country') according to Aslet (1982), 'function, or gave the illusion that they functioned, as the centres of landed estates and had attached to them an estate or at least a home farm, and also the offices, outbuildings and lodges associated with a substantial house' (p. 2).

Aslet lists just over 200 of such 'country houses' built between 1890 and 1930 and though the economic basis for their construction is not given in all cases, it is clear that a substantial proportion represent the investment of overseas-derived profits. As to their location, the largest majority were built in south-eastern counties: Surrey (26), Sussex (18), Hampshire (14), Norfolk (10), Kent (9), Gloucestershire (8), Buckinghamshire (7), Oxfordshire (6), and Suffolk (6). The main exception in this south-east concentration is Yorkshire (7).

Other counties with two or more country houses constructed in the same period include:

Two	*Three*	*Four*	*Five*
Argyllshire	Worcestershire	Hertfordshire	Dorset
Invernesshire	Wiltshire	Devonshire	
Aberdeenshire	Westmoreland		
Leicestershire	Cambridgeshire		
Ayrshire	Berkshire		

Lancashire	Cheshire
Lincolnshire	Northumberland
Shropshire	Glamorgan
Northamptonshire	East Lothian
Berwickshire	

Typical of those built at the height of British hegemony in the world-economy and the largely overseas (often colonial) profits that funded them are listed in Table 7.3.

Table 7.3 The international basis of country house building in Britain, 1890–1930

House	Location by county	Status or business of building patron	Date of building
Castle Drogo	Devon	Co-founder of Home and Colonial Stores	1910–30
Caythorpe Court	Lincolnshire	Banker and brewer	1901
Chelwood Manor	Sussex	International railway contractor	1904
Clock House	Sussex	Partner, Glynn Mills Bank	1913–14
Coldharbour Wood	Hampshire	Chairman, Peninsular & Orient Steamship Co.	1889–1916
Crooksbury	Surrey	Calcutta-based business	1898
Flint House	Oxfordshire	Wholesale grocer	1913
Gledstone Hall	Yorkshire	Cotton broker	1906
Harcombe Chudleigh	Devon	Tobacco manufacturer	1913
Heathcote	Yorkshire	Cotton broker	1906
Kelling Hall	Norfolk	Director-General of Royal Dutch Petroleum	1912–13
Luton Hoo	Bedfordshire	Diamond business	1903
Pangbourne Tower	Berkshire	Shipbuilder	1898
Pickenham Hall	Norfolk	Banker	1902
Skibo Castle	Sutherland	American steel millionaire (Carnegie)	1899–1903
Stowell Hill	Somerset	Meat importer (Lord Vestey)	1927
Tigbourne Court	Surrey	Chairman. Prudential Assurance	1899
Wilsford Manor	Wiltshire	Chemical manufacturer	1904
Witley Park	Surrey	Financier	1890
Woodfalls	Hampshire	Founder of Imperial Chemical Industries	1930

Although the definition of what qualifies as a 'historic country house' in Britain is problematic, Cornforth (1974: 4) suggests a figure of 1,500 'notable country houses' and gives other sources indicating 'probably less than 2000'.

SELECT BIBLIOGRAPHY

Abu-Lughod, J. (1965) 'Tale of two cities: the origins of modern Cairo', *Comparative Studies in Society and History* 7: 429–57.

Abu-Lughod, J. (1971) *Cairo: 1001 Years of the City Victorious*, Princeton, NJ: Princeton University Press.

Abu-Lughod, J. (1975) 'Moroccan cities: apartheid and the serendipity of conservation', in I. Abu-Lughod (ed.) *African Themes*, Northwestern University Studies in honour of G.M. Carter, Evanston, pp. 77–111.

Abu-Lughod, J. (1976) 'Developments in North African urbanism. The process of decolonization', in B.J.L. Berry (ed.) *Urbanization and Counterurbanization*, Beverly Hills, CA and London: Sage, pp. 191–211.

Abu-Lughod, J. (1978) 'Dependent urbanism and decolonization: the Moroccan case', *Arab Studies Quarterly* 1: 49–66.

Abu-Lughod, J. (1980) *Rabat. Urban Apartheid in Morocco*, Princeton, NJ: Princeton University Press.

Abu-Lughod, J. (1984) 'Culture, "modes of production" and the changing nature of cities in the Arab world', in J. Agnew, J. Mercer, and D. Sopher (eds) *The City in Cultural Context*, London: Allen & Unwin, pp. 44–117.

Abu-Lughod, J. and Hay, R. (eds) (1977) *Third World Urbanization*, Chicago: Maaroufa Press.

Acquah, I. (1958) *Accra Survey*, London: University of London.

Adu Boahen, A. (1987) *African Perspectives on Colonialism*, Baltimore, MD: Johns Hopkins Press.

Agnew, J. (1981) 'Home ownership and identity in capitalist societies', in J.S. Duncan (ed.) *Housing and Identity. Cross-cultural Perspectives*, London: Croom Helm, pp. 60–97.

Aiken, M. and Castells, M. (1977) 'New trends in urban studies', *Comparative Urban Research* 4 (2, 3): 7–10.

Alavi, H. (1980) 'India: transition from feudalism to colonial capitalism', *Journal of Contemporary Asia* 10: 359–99.

Alcock, A.E.S. and Richards, H. (1953) *How to Plan Your Village. A Handbook for Villages in Tropical Countries*, London: Longmans, Green.

Amin, S. (1974) *Accumulation on a World Scale. A Critique of the Theory of*

Underdevelopment, New York: Monthly Review Press.

Ardener, S. (1981) *Women and Space. Ground Rules and Social Maps*, London: Croom Helm.

Armstrong, W. and McGee, T.G. (1985) *Theatres of Accumulation. Studies in Asian and Latin American Urbanization*, London: Methuen.

Asad, T. (1973) *Anthropology and the Colonial Encounter*, London: Ithaca Press.

Aslet, C. (1982) *The Last Country Houses*, New Haven, CT: Yale University Press.

Atkinson, G.A. (1953) 'British architects in the tropics', *Architectural Association Journal* 69 (773): 7–21.

Atkinson, G.A. (1959) 'Recent advances in low-cost housing in tropical areas', Proceedings, Sixth International Congress of Tropical Medicine and Malaria.

Balandier, G. (1951) 'The colonial situation: A theoretical approach', in I. Wallerstein (ed.) (1966) *Social Change: The Colonial Situation*, New York: Wiley, pp. 34–61.

Ball, M. (1983) *Housing Policy and Economic Power. The Political Economy of Owner Occupation*, London: Methuen.

Ballhatchet, K. (1979) *Race, Sex and Class under the Raj*, London: Blackwell.

Ballhatchet, K. and Harrison, J. (1980) *The City in South Asia*, London: Curzon Press.

Barratt Brown, M. (1978) *The Economics of Imperialism*, Harmondsworth: Penguin.

Bartlett Summer School (1979–) *The Production of the Built Environment*, Bartlett School of Architecture and Planning, University College, London: UCL.

Basu, D.K. (ed.) (1985) *The Rise and Growth of Colonial Port Cities in Asia*, Santa Cruz: Center for South Pacific Studies, University of California (1st edn, 1979).

Bayly, C. (1983) *Rulers, Townsmen and Bazaars: North Indian Society in the Age of British Expansion, 1770–1870*, Cambridge.

Becker, P.G., Frieden, J., Schatz, S.B., and Sklar, R.L. (1987) *Post-Imperialism. International Capital and Development in the Late Twentieth Century*, Boulder CO and London: Rienner.

Bellam, M.E.P. (1970) 'The colonial city: Honaira, a Pacific island's case study', *Pacific Viewpoint* 11 (1): 66–96.

Benevolo, L. (1980) *The History of the City*, London: Scolar Press.

Benoit-Levy, G. (1930) *Maisons de Campagne sans Etage et Bungalows*, Bruxelles and Paris: Salmain & Fils.

Bergesen, A. (ed.) (1980) *Studies of the Modern World System*, London: Sage.

Bergeson, A. and Schoenberg, R. (1980) 'Long waves of colonial expansion and contraction, 1415–1969', in A. Bergeson (ed.) *Studies of the Modern World System*, London: Sage.

Berry, B.J.L. (1976) 'The counter-urbanization process: urban America since 1970', in *Urbanization and Counter-urbanization*, Beverly Hills, CA and London: Sage, pp. 17–30.

Betting, W. and Vriend, J. (1958) *Bungalows: Deutschland, England, Italien, Holland, Belgien, Danemark*, Darmstadt: Die Planung Verlag, Muller-Welborn.

Betts, R.F. (1969) 'The problem of the medina in the urban planning of Dakar, Senegal', *African Urban History* September.

Betts, R.F. (n.d.) 'The architecture of French African Empire. A neglected history', Department of History, University of Kentucky, Lexington (unpublished).

Blau, J.R. (1984) *Architects and Firms. A Sociological Perspective on Architectural Practice*, Cambridge, MA: MIT Press.

Blussé, L. (1985) 'An insane administration and an unsanitary town; The Dutch East India Company and Batavia, 1619–1799', in R. Ross and G. Telkamp (eds) *Colonial Cities*, Boston, Lancaster, Dordrecht: Martinus Nijhoff, pp. 65–86.

Boralévi, A. (1980) 'Le "Citta' dell' impero": Urbanistica fascista in Etiopia, 1936–41', in A. Moini (ed.) *Urbanistica Fascista*, Milan: Cinta.

Braudel, F. (1984) *The Perspective of the World*, London: Fontana.

Braudel, F. (1985) *The Structures of Everyday Life*, London: Fontana.

Briggs, R.A. (1891) *Bungalows and Country Residences*, London: Batsford.

Brittan, A. (1977) *The Privatized World*, London: Routledge & Kegan Paul.

Browning, H. and Roberts, B. (1980) 'Urbanisation, sectoral transformation and the utilization of labour in Latin America', *Comparative Urban Research* 8 (1): 86–104.

Brunn, S.D. and Williams, J.F. (1983) *Cities of the World: World Regional Urban Development*, New York: Harper and Row.

Brush, J.E. (1970) 'The growth of the Presidency Towns', in R.G. Fox (ed.) *Urban India: Society, Space and Image*, Durham, NC: Duke University Press, pp. 91–114.

Brush, J.E. (1974) 'Spatial patterns of population in Indian cities', in D.J. Dwyer (ed.) *The City in the Third World*, London and New York: Barnes & Noble, pp. 105–32.

Burgess, R. (1978) 'Petty commodity housing or dweller control? A critique of John Turner's views on housing policy', *World Development* 6: 1105–34.

Caffyn, L. (1986) *Workers' Housing in West Yorkshire 1750–1920*, London: HMSO.

Carlstein, T., Parkes, D., and Thrift, N. (eds) (1978) *Timing Space and Spacing Time* (3 vols), London: Edward Arnold.

Carr, M.C. (1982) 'The development and character of a metropolitan suburb: Bexley, Kent', in F.M.L. Thompson (ed.) *The Rise of Suburbia*, Leicester: Leicester University Press, pp. 211–69.

Carswell, J.W. and Saile, D.G. (eds) (1986) *Built Form and Culture Research*, Conference Proceedings, University of Kansas, Department of Architecture.

Castells, M. (1972) *Imperialismo y Urbanization en America Latina*, Barcelona: Editorial Gustava Gili.

Castells, M. (1977) *The Urban Question*, London: Edward Arnold (first published as *La Question Urbaine*, Paris: Francis Maspero (1972).

Cavanaugh, J. and Clairmonte, F. (1984) *Transnational Corporations and Global Markets*, Washington, DC: Institute for Policy Studies.

Chaichian, M.A. (1988) 'The effects of world capitalist economy on urbanization in Egypt, 1800–1970', *International Journal of Middle Eastern Studies* 20: 23–43.

Chapman, S.D. (ed.) (1971) *The History of Working Class Housing*, Newton Abbot: David & Charles.

Chase-Dunn, C. (1985) 'The system of world cities, AD 800–1975', in M. Timberlake (ed.) *Urbanization in the World-Economy*, London: Academic Press.

Checkoway, B. (1980) 'Large builders, federal housing programmes and postwar suburbanisation', *International Journal of Urban and Regional Research* 4: 21–45.

Cherry, G. (ed.) (1980) *Shaping an Urban World*, London: Mansell.

Chilcote, R.H. (1984) *Theories of Development and Underdevelopment*, Boulder, CO: Westview Press.

Chirot, D. (1986) *Social Change in the Modern Era*, New York and London: Harcourt Brace Jovanovich.

Christopher, A.J. (1988) *The British Empire at its Zenith*, London: Croom Helm.

Clarke, C.G. (1975) *Kingston Jamaica. Urban Development and Social Change, 1692–1962*, Berkeley, CA: University of California Press.

Clarke, C.G. (1985) 'A Caribbean creole capital: Kingston, Jamaica, 1692–1938', in R. Ross and G.J. Telkamp (eds) *Colonial Cities*, Boston, Lancaster, Dordrecht: Martinus Nijhoff, pp. 153–70.

Cohen, R.B. (1981) 'The new international division of labor, multinational corporations and urban hierarchy' in M. Dear and A.J. Scott (eds) *Urbanization and Urban Planning in Capitalist Society*, London: Methuen, pp. 287–315.

Cohen, R. (1987) *The New Helots. Migrants in the International Division of Labor*, Aldershot: Gower Publishing.

Collins, J. (1980) 'Lusaka: urban planning in a British colony' in G.E. Cherry (ed.) *Shaping an Urban World*, London: Mansell.

Colonial Building Notes, 1950–7 (later, *Overseas Building Notes*, 1959–present), Building Research Station, Garston, Watford.

Conlon, F. (1981) 'Working class housing in colonial Bombay', Proceedings of the Seventh European Conference on South Asian Studies, School of Oriental and African Studies, London, July.

Cooke, P. (ed.) (1986) *Global Restructuring, Local Response*, Redhill, Surrey: Schools Publishing Co.

Cooke, P. (ed.) (1989) *Localities*, London: Hutchinson (forthcoming).

Cooper, I. (1982) 'Comfort theory and practice. Barriers to the conservation of energy by building occupants', *Applied Energy* 11 (14): 243–69.

Corbridge, S. (1986) *Capitalist World Development*, London: Macmillan.

Cornforth, J. (1974) *Country Houses in Britain. Can they Survive?* London: *Country Life* for British Tourist Authority.

Cross, M. (1979) *Urbanisation and Urban Growth in the Caribbean*, Cambridge: Cambridge University Press.

Curtin, P. (1985) 'Medical knowledge and urban planning in tropical Africa', *American Historical Review* 90 (3): 594–613.

Darley, G. (1985) 'Existential cities', in R. Fermor-Hesketh (ed.) *Architecture of the British Empire*, London: Weidenfeld & Nicolson, pp. 74–103.

Daunton, M.J. (1983) *House and Home in the Victorian City*, London: Edward Arnold.

Davies, D.H. (1969) 'Lusaka, Zambia: some town-planning problems in an African capital city at Independence', *Zambian Urban Studies*, Lusaka: Zambian Institute of Social Research.

Dear, M. and Scott, A.J. (eds) (1981) *Urbanization and Urban Planning in Capitalist Society*, London: Methuen.

de Bruyne, G.A. (1985) 'The colonial city and the post-colonial world', in R. Ross and G.J. Telkamp (eds), *Colonial Cities*, Boston, Lancaster, Dordrecht: Martinus Nijhoff, pp. 231–44.

Denyer, S. (1978) *African Traditional Architecture*, London: Heinemann.

Dethier, J. (1973) 'Evolution of concepts of housing, urbanism and country planning in a developing country: Morocco', in L.C. Brown (ed.) *From Madina to Metropolis*, Princeton: Darwin Press, 197–243.

de Vries, J. (1983) *European Urbanization, 1500–1800*, London: Methuen.

Dossal, M. (1989) 'Colonial urban planning in Bombay, 1860–1880', *International Journal of Urban and Regional Research* 13, 1 (in press).

Douglas, M. (ed.) (1979) *Rules and Meanings*, Harmondsworth: Penguin Books.

Doxtater, D. (1984) 'Spatial opposition in non-discursive expression; architecture as ritual process', *Canadian Journal of Anthropology* 4: 1–17.

Drakakis-Smith, D. (1987) *The Third World City*, New York: Methuen.

Duclos, D. (1981) 'The capitalist state and the management of time', in M. Harloe and E. Lebas (eds) *City, Class and Capital*, London: Edward Arnold.

Duncan, N.G. (1981) 'Home ownership and social theory', in J.S. Duncan (ed.) *Housing and Identity. Cross-cultural Perspectives*, London: Croom Helm, pp. 98–134.

Dunning, J.H. (1983) 'Changes in the level and structure of international production: the last one hundred years', in M. Casson (ed.) *The Growth of International Business*, London: Allen & Unwin, pp. 84–139.

Dwyer, D.J. (ed.) (1974) *The City in the Third World*, London and New York: Barnes & Noble.

Edelstein, M. (1981) 'Foreign investment in empire', in R. Floud and D. McCloskey (eds) *The Economic History of Britain since 1700, part 2, 1860–1970s*, Cambridge: Cambridge University Press, pp. 70–98.

Elliot, M. (1986) *Heartbeat London*, London: Firethorn Press.

Emerson, R. (1968) 'Colonialism', in *International Encyclopaedia of Social Sciences*, New York: Macmillan.

Evans, R. (1983) *The Fabrication of Virtue: English Prison Architecture, 1750–1840*, Cambridge: Cambridge University Press.

Feagin, J.R. and Smith, M.P. (1987) 'Cities and the new international division of labour. An overview', in M.P. Smith and J.R. Feagin (eds) *The Capitalist City*, Oxford: Blackwell.

Fermor-Hesketh, R. (ed.) (1985) *Architecture of the British Empire*, London: Weidenfeld & Nicolson.

Fetter, B. (1976) *The Creation of Elizabethville, 1910–40*, Stanford: Hoover Institute Press.

Fielding, A.J. (1982) 'Counterurbanisation in Western Europe', *Progress in Planning* 17: 1–52.

Floud, R. and McCloskey, D. (1981) *The Economic History of Britain since 1700, part 2, 1860–1970s*, Cambridge: Cambridge University Press.

Foglesong, R.E. (1986) *Planning the Capitalist City: the Colonial Era to the 1920s*, Princeton, NJ: Princeton University Press.

Forbes, D. and Thrift, N.J. (eds) (1987) *The Socialist Third World. Urban, Development and Territorial Planning*, Oxford: Blackwell.

Forty, A. (1980) 'The modern hospital in England and France: the social and medical uses of architecture', in A.D. King (ed.) *Buildings and Society: Essays on the Social Development of the Built Environment*, London: Routledge & Kegan Paul, pp. 61–93.

Foster, J. (1974) *Class Struggle and the Industrial Revolution*, London: Methuen.

Foucault, M. (1973) *Madness and Civilization*, New York: Vintage Books.

Foucault, M. (1979) *Discipline and Punish. The Birth of the Prison*, New York: Vintage Books.

Fraser, D. (1980) *A History of Modern Leeds*, Manchester: University of Manchester Press.

Friedmann, J. (1986) 'The world city hypothesis', *Development and Change* 17 (1): 69–83.

Friedmann, J. and Wolff, G. (1982) 'World city formation: an agenda for research and action', *International Journal of Urban and Regional Research* 6: 309–44.

Friedmann, J. and Wulff, R. (1976) *The Urban Transition: Comparative Studies of Newly Industrializing Societies*, London: Arnold.

Frenkel, S. and Western, J. (1988) 'Pretext or prophylaxis? Racial segregation and malarial mosquitos in a British tropical colony', *Annals of the American Association of Geographers* 78,2.

Frobel, F., Heinrichs, J., and Kreye, O. (1980) *The New International Division of Labour*, Cambridge: Cambridge University Press.

Furedy, C. (1979) 'The development of modern elite retailing in Calcutta, 1890–1920', *Indian Economic and Social History Review* 16: 377–92.

Gadgil, D.R. (1973) *The Industrial Evolution of India in Recent Times, 1860–1939*, Bombay: Oxford University Press.

Gamble, A. (1985) *Britain in Decline*, London: Macmillan.

Gardner-Medwin, R. *et al.* (1948) 'Recent planning developments in the colonies', *RIBA Journal* 55, February.

Gershuny, J. (1978) *After Industrial Society? The Emerging Self-Service Economy*, London: Macmillan.

Giddens, A. (1984) *The Constitution of Society*, Cambridge: Polity Press.

Giese, E. (1979) 'Transformation of Islamic cities in Soviet Middle Asia into socialist cities', in R.A. French and F.E.I. Hamilton (eds) *The Socialist City. Spatial Structure and Urban Policy*, Chichester: Wiley.

Gill, G. (n.d. but *c*. 1901) *The British Colonies, Dependencies and Protectorates*, London: George Gill and Sons.

Ginsburg, N. (1965) 'Urban geography in non-Western areas', in P.M. Hauser and L.F. Schnore (eds) *The Study of Urbanization*, New York: Wiley, pp. 311–46.

Girouard, M. (1971) *The Victorian Country House*, London: Country Life.

Glickman, N.J. (1987) 'Cities and the international division of labour', in M.P. Smith and J.R. Feagin (eds) *The Capitalist City*, Oxford: Blackwell, pp. 66–86.

Gordon, D. (1984) 'Capitalist development and the history of American cities', in W.K. Tabb and L. Sawers (eds) *Marxism and the Metropolis: New Perspectives in Urban Political Economy*, Oxford: Oxford University Press.

Gregory, D. and Urry, J. (1985) *Social Relations and Spatial Structures*, New York: Macmillan.

Gugler, J. (ed.) (1970) *Urban Growth in Sub-Saharan Africa*, Kampala: Makere Institute of Social Research.

Gutkind, P.C. (1981) 'Review of King, A.D.: Colonial Urban Development', *International Journal of Urban and Regional Research* 5: 290–1.

Hall, P., Gracey, H., Drewett, R., and Thomas, R. (1973) *The Containment of Urban England* (2 vols), London: Allen & Unwin.

Hallowell, I. (1967) *Culture and Experience*, Philadelphia, PA: University of Pennsylvania Press.

Hardoy, J.E. and Satterthwaite, D. (eds) (1981) *Shelter, Need and Response: Housing, Land and Settlement in Seventeen Third World Nations*, Chichester: John Wiley.

Harley, C.K. and McCloskey, D.N. (1981) 'Foreign trade: competition and the expanding international economy', in R. Floud and D. McCloskey (eds) *The Economic History of Britain since 1700, part 2, 1860-1970s*, Cambridge: Cambridge University Press, pp. 50–70.

Harloe, M. (ed.) (1977) *Captive Cities. Studies in the Political Economy of Cities and Regions*, London: Edward Arnold.

Harloe, M. (1987) 'Editorial introduction', *International Journal of Urban and Regional Research* 11: 1.

Harloe, M. and Lebas, E. (eds) (1981) *City, Class and Capital. New Developments in the Political Economy of Cities and Regions*, London: Edward Arnold.

Harvey, D. (1973) *Social Justice and the City*, London: Edward Arnold.

Harvey, D. (1975) Review of B.J.L. Berry, *The Human Consequences of Urbanisation*, London, 1973, in *Annals of the American Association of Geographers* 65: 99–103.

Harvey, D. (1985) *The Urbanization of Capital*, London: Edward Arnold.

Harvey, D. (1987) 'Flexible accumulation through urbanization:

reflections on "post-modernism" in the American city', paper presented to the Sixth Urban Change and Conflict Conference, University of Kent, September.

Held, D. (1980) *Introduction to Critical Theory. Horkheimer to Habermas*, London: Hutchinson.

Henderson, J. and Castells, M. (eds) (1987) *Global Restructuring and Territorial Development*, London and Beverly Hills: Sage.

Herington, J. (1984) *The Outer City*, London: Harper & Row.

Hill, R.C. (1977) 'Capital accumulation and urbanization in the United States', *Comparative Urban Research* 14 (2, 3).

Hill, R.C. (1984) 'Urban political economy' in M.P. Smith (ed.) *Cities in Transformation*, Beverly Hills, CA and London: Sage, pp. 123–38.

Hines, T. (1972) 'The imperial facade: D.H. Burnham and American architectural planning in the Philippines', *Pacific Historical Review* 61: 35–53.

Hitchcock, H.R. and Johnson, J. (1966) *The International Style*, New York: W.W. Norton.

Hobsbawm, E.J. (1969) *Industry and Empire*, Harmondsworth: Penguin.

Hobsbawm, E.J. (1975) *The Age of Capital*, London: Weidenfeld & Nicolson.

Hobson, J.A. (1948). *Imperialism. A Study*, London: Allen & Unwin.

Holzman, J.M. (1926) *The Nabobs in England, 1760-5*, New York: Columbia University Press.

Home, R.K. (1983) 'Town planning, segregation and indirect rule in colonial Nigeria', *Third World Planning Review* 5 (2): 165–75.

Hopkins, A.G. (1980) 'Property rights and empire building: the British annexation of Lagos, 1861', *Journal of Economic History* 40: 777–98.

Horvath, R.V. (1969) 'In search of a theory of urbanization: notes on the colonial city', *East Lakes Geographer* 5: 68–82.

Horvath, R.V. (1972) 'A definition of colonialism', *Current Anthropology* 13(1): 45–57.

Hull, R.W. (1976) *African Cities and Towns before the European Conquest*, New York: Horton.

Ingham, G. (1984) *Capitalism Divided*, Longon: Macmillan.

Irving, R. (ed.) (1985) *The History and Design of the Australian House*, Melbourne and Oxford: Oxford University Press.

Irving, R.G. (1983) *Indian Summer. Lutyens, Baker and Imperial Delhi*, New Haven: Yale University Press.

Jackson, A.A. (1973) *Semi-detached London*, London: Allen & Unwin.

Jameson, F. (1985) 'Postmodernism and consumer society', in H. Foster (ed.) *Postmodern Culture*, London: Pluto Press, pp. 111–25.

Johnston, R.J. (1980) *City and Society. An Outline for Urban Geography*, Harmondsworth: Penguin.

Kanyeihamba, G.W. (1980) 'The impact of the received law on planning and development in Anglo-phonic Africa', *International Journal of Urban and Regional Research* 4: 62–84.

Karasch, M. (1985) 'Rio de Janeiro: from colonial town to imperial capital', in R. Ross and G.J. Telkamp (eds), *Colonial Cities*, Boston, Lancaster, Dordrecht: Martinus Nijhoff, pp. 123–54.

Kaye, B. (1960) *The Development of the Architectural Profession in Britain*, London: Allen & Unwin.

Kemeny, J. (1981) *The Myth of Home Ownership*, London: Routledge & Kegan Paul.

Kentor, J. (1985) 'Economic development and the world division of labour', in M. Timberlake (ed.) *Urbanization in the World-Economy*, London: Academic Press, pp. 25–40.

King A.D. (1976) *Colonial Urban Development: Culture, Social Power and Environment*, London: Routledge & Kegan Paul.

King, A.D. (1977) 'Exporting planning: the colonial and neo-colonial experience', *Urbanism Past and Present* 5 (Winter): 12–22.

King, A.D. (ed.) (1980) 'A space for time and a time for space: the social production of the vacation house', in A.D. King (ed.) *Buildings and Society: Essays on the Social Development of the Built Environment*, London: Routledge & Kegan Paul, pp. 193–227 (2nd edn. 1984).

King, A.D. (1982) 'Colonial architecture and urban development; the reconversion of colonial typologies', *Lotus International* 34: 46–59.

King, A.D. (1983) ' "The world economy is everywhere": urban history and the world-system', *Urban History Yearbook, 1983*, Leicester University Press.

King, A.D. (1984) *The Bungalow: the Production of a Global Culture*, London: Routledge & Kegan Paul.

King, A.D. (1985) 'Colonial Cities: global pivots of change', in R. Ross and G. Telkamp (eds) *Colonial Cities*, Dordrecht, Boston, Lancaster: Martinus Nijhoff for the Leiden University Press.

King, A.D. (1986a) 'Margins, peripheries and divisions of labour: UK urbanism and the world-economy', in D. Hardy (ed.) *On the Margins: Marginal Space and Marginal Economies*, Middlesex Polytechnic Planning Paper, no. 17.

King, A.D. (1987) 'Cultural production and reproduction: the political economy of societies and their built environment ', in D. Cantor, M. Krampen, and D. Stea (eds) *Ethnoscapes: Transcultural Studies in Action and Place*, London: Gower.

King A.D. (1989) 'Colonialism, urbanism, and the capitalist world-economy: an introduction', *International Journal of Urban and Regional Research* 13 (1): 1–18.

King, A.D. (1990) *Global Cities: Post-imperialism and the Internationalization of London*, London: Routledge.

King, A.D. (ed.) (1991) *Culture, Globalization and the World-System: Contemporary Conditions for the Representation of Identity*, Current Debates in Art History 3, State University of New York at Binghamton (forthcoming).

Kira, A. (1977) *The Bathroom*, Harmondsworth: Penguin.

Kirk, G. (1980) *Urban Planning in Capitalist Society*, London: Croom Helm.

Knox, P. (ed.) (1988) *The Design Professions and the Built Environment*, London: Croom Helm.

Kooiman, D. (1985) 'Bombay: from fishing village to colonial port city, 1662–1947', in R. Ross and G.J. Telkam (eds) *Colonial Cities*,

Boston, Lancaster, Dordrecht: Martinus Nijhoff.

Kosambi, M. (1985) 'Commerce, conquest and the colonial city: the role of locational factors in the rise of Bombay', *Economic and Political Weekly* 5: 31–37, January.

Kumar, K. (ed.) (1980) *Transnational Enterprises. Their Impact on Third World Societies and Cultures*, Boulder, CO: Westview Press.

Lampard, E.E. (1986) 'The New York metropolis in transformation: history and prospect', in H.J. Ewers, J.B. Goddard, and H. Matzerath (eds) *The Future of the Metropolis*, Berlin and New York: de Gruyter.

Lancaster, C. (1985) *The American Bungalow*, New York: Abbeville Press.

Langlands, B. (1969) 'Perspective on urban planning for Uganda', in M. Safier and B.W. Langlands (eds) *Perspectives on Urban Planning for Uganda*, Uganda: Department of Geography, Makerere University College.

Lash, S. and Urry, J. (1987) *The End of Organized Capitalism*, Cambridge: Polity Press.

Learmouth, A.T.L. and Spate, O.U.K. (1965) *India: A Regional Geography*, London: Methuen.

Lefebvre, H. (1970) *La Revolution Urbain*, Paris: Gaillimard.

Lewandowski, S.J. (1977) 'Changing form and function in the ceremonial and colonial port city in India', *Modern Asian Studies* 11: 183–213.

Lewcock, R. (1963) *Early Nineteenth Century Architecture in South Africa*, Cape Town: Hobbema.

Lewcock, R. (1979) Review of King, *Colonial Urban Development (1976)*, in *Modern Asian Studies* 13: 164–7.

Little, A. (1974) *Urbanisation as a Social Process*, London: Routledge & Kegan Paul.

Lloyd, P. (1979) *Slums of Hope. Shanty Towns of the Third World*, Harmondsworth: Pelican.

Lock, Max, and Partners (1967) *Survey and plan for Kaduna, Northern Nigeria*, London: Faber & Faber.

Lockard, C.A. (1976) 'The early development of Kuching; 1820–1857', *Journal of the Malaysian Branch of the Royal Asiatic Society* 49: 107–26.

Lojkine, J. (1976) 'Contribution to a Marxist theory of capitalist urbanisation', in C. Pickvance (ed.) *Urban Sociology, Critical Essays*, London: Methuen, pp. 119–46.

Lotus Internatinal (1980) 'Architecture and colonialism', 26, Milan.

Lowder, S. (1986) *Inside Third World Cities*, London: Croom Helm.

Lubeck, P. and Walton, J. (1979) 'Urban class and conflict in Africa and Latin America', *International Journal of Urban and Regional Research* 3: 2–29.

Mabogunje, A.L. (1980) *The Development Process: a Spatial Perspective*, London: Hutchinson.

Mabogunje, A.L., Hardoy, J.E., and Misra, R.P. (1978) *Shelter Provision in Developing Countries: The Influence of Standards and Criteria*, Chichester: Wiley.

Maclean, Charter, (1982) 'Kuwait. A place of architectural pilgrimage?', *The Times*, 1 June 1982, p. viii.

Makler, H., Martinelli, A., and Smelser, N. (eds) (1982) *The New International Economy*, Beverly Hills, CA and London: Sage.

Manzoor Alam, S. (1972) *Metropolitan Hyderabad and its Region. A Strategy for Development*, London: Asia Publishing House.

Marshall, P. (1985) 'Eighteenth century Calcutta', in R. Ross and G.J. Telkamp (eds) *Colonial Cities*, Boston, Lancaster, Dordrecht: Martinus Nijhoff.

Massey, D. (1979) 'In what sense a regional problem?', *Regional Studies* 13: 233–43.

Massey, D. (1984) *Spatial Divisions of Labour*, London: Macmillan.

Massey, D. (1986) 'The legacy lingers on: the impact of Britain's historical international role on its internal geography', in R. Martin and R. Rowthorn (eds) *Deindustrialization and the British Economy*, London: Macmillan.

Massey, D. and Meegan, R. (1982) *The Anatomy of Job Loss*, London: Methuen.

McCauslen, P. (1981) 'The legal environment of planned urban growth', Occasional Paper, Development Planning Unit, University College, 9 Endesleigh Gardens, London WC1 0ED.

McGee, T.G. (1967) *The South East Asian City*, London: Bell.

Metcalfe, T.R. (1989) *An Imperial Vision. Indian Architecture and Britain's Raj*, Berkeley, CA: University of California Press.

Mingione, E. (1981) *Social Conflict and the City*, London: Edward Arnold.

Mitchell, J.C. (1966) 'Theoretical orientations in African urban studies', in M. Banton (ed.) *The Social Anthropology of Complex Societies*, London: Tavistock.

Morris, J. (1985) 'In quest of imperial style', in R. Fermor-Hesketh (ed.) *Architecture and the British Empire*, London: Weidenfeld & Nicolson.

Mullins, P. (1981) 'Theoretical perspectives on Australian urbanization. Material components in the production of Australian labour power', *Australian and New Zealand Journal of Sociology* 17: 65–76.

Murphy, R. (1969) 'Traditionalism and colonialism: changing urban roles in Asia', *Journal of Asian Studies*, 29 (1): 67–84.

Muthesius, S. (1983) *The English Terrace House*, New Haven, CT: Yale University Press.

Naqvi, H.K. (1968) *Urban Centres and Industries in Upper India 1553–1803*, Bombay: Asia Publishing House.

Neild, S.M. (1979) 'Colonial urbanism: the development of Madras city in the eighteenth and nineteenth centuries', *Modern Asian Studies* 13: 217–46.

NHBC (1984) 'Housing for sale to the elderly', National House Building Council Second Report prepared for the Housing Research Foundation; copy available from National House Building Council, 58 Portland Place, London W1.

Nilsson, S. (1973) *The New Capitals of India, Pakistan and Bangladesh*, Lund: Scandinavian Institute of South Asian Studies.

Oldenburg, V.T. (1984) *The Making of Colonial Lucknow, 1856–77*, Princeton: Princeton University Press.

Oliver, P. (1969) *Shelter and Society*, London: Barrie and Jenkins.

Oliver, P. (ed.) (1971) *Shelter in Africa*, London: Heinemann.

Oliver, P. (1987) *Dwellings. The House Across the World*, London: Phaidon.

Omvedt, G. (1973) 'A definition of colonialism', *The Insurgent Sociologist*, pp. 1–24, Spring.

O'Neill, S. (1982) 'Time-sharing. Its Implications for Building Development', Final Year Project, Department of Building Technology, Brunel University, Uxbridge.

Oosterhoff, J.L. (1985) 'Zeelandia. A Dutch colonial city on Formosa, 1624–1662', in R. Ross and G. Telkamp (eds) *Colonial Cities*, Boston, Lancaster, Dordrecht: Martinus Nijhoff, 51–64.

Oxborough, I. (1979) *Theories of Underdevelopment*, London: Macmillan.

Pahl, R. (1970) *Patterns of Urban Life*, London: Longman.

Pahl, R. (1984) 'The restructuring of capital, the local political economy and household work strategies', in D. Gregory and J. Urry (eds) *Social Relations and Spatial Structures*, London: Macmillan, pp. 242–64.

Pahl, R.E., Flynn, R., and Buck, N.R. (1983) *Structures and Processes of Urban Life*, London: Longman.

Palat, R., Barr, K., Matson, J., Bahl, V., and Ahmad, N. (1986) 'The incorporation and peripheralization of South Asia, 1600–1950', *Review* 10 (1): 171–208.

Pang, E.S. (1983) 'Buenos Aires and the Argentine economy in world perspective, 1776–1930', *Journal of Urban History* 9 (3): 365–82.

Parkin, D. (1972) *Town and Country in Central and Eastern Africa*, London: Oxford University Press.

Parry, L.J. (1965) *Building Cycles and Britain's Growth*, London: Macmillan.

Payne, G. (ed.) (1984) *Low Income Housing in the Third World*, Chichester: Wiley.

Pickvance, C. (1976) *Urban Sociology. Critical Essays*, London: Methuen.

Portes, A. and Walton, J. (1981) *Labor, Class and the International System*, London: Academic Press.

Power, J. (1979) *Migrant Workers in Western Europe and the United States*, Oxford: Pergamon.

Pradilla, E. and Jimenez, C. (1985) 'Architecture, urbanism and neocolonial dependence', in R. Bromley (ed.) *Planning for Small Enterprises*, Oxford: Pergamon, pp. 191–202.

Pred, A. (1977) *City-Systems in Advanced Economies. Past Growth, Present Processes and Future Development Options*, New York: Wiley.

Pred, A. (1980) *Urban Growth and City Systems in the United States, 1840–1860*, Cambridge, MA: Wiley.

Preston, P.W. (1986) *Making Sense of Development*, London: Routledge & Kegan Paul.

Prior, L. (1988) 'The architecture of the hospital: a study of spatial organisation and medical knowledge', *British Journal of Sociology* 39 (1): 86–113.

Property International ('The magazine covering leisure and investment real estate around the world') (1985) vol. 1, 2, Falcon Publishing Europe Ltd, 27–29 Queen Anne St, London W1V.

Rabinow, P. (1989a) 'Modernity and difference: French colonial planning in Morocco', *International Journal of Urban and Regional Research*, 13 (1).

Rabinow, P. (1989b) *French Modern. Norms and Forms of Missionary and Didactic Pathos*, Cambridge, MA: MIT Press (forthcoming).

Rapoport, A. (1969) *House Form and Culture*, Englewood Cliffs, NJ: Prentice-Hall.

Rapoport, A. (ed.) (1976) *The Mutual Interaction of People and Their Built Environment: A Cross-Cultural Perspective*, The Hague: Mouton.

Rapoport, A. and Watson, N. (1972) 'Cultural variability in physical standards', in R. Gutman (ed.) *People and Buildings*, New York: Basic Books.

Rayfield, J.R. (1974) 'Theories of urbanization and the colonial city in West Africa', *Africa*, 44: 163–85.

Redfield, R. and Singer, M.S. (1954) 'The cultural role of cities', *Economic Development and Cultural Change* 3: 53–73.

Reed, R.R. (1976) *City of Pines. The Origins of Baguio as a Colonial Hill Station*, Berkeley, CA: Center for South and South-East Asian Studies, University of California.

Reed, R.R (1978) *Colonial Manila. The Context of Hispanic Urbanism and the Process of Morphogenesis*, Berkeley and Los Angeles: University of California Publications in Geography, vol. 22.

Rees, G. and Lambert, J. (1985) *Cities in Crisis. The Political Economy of Urban Development in Post-War Britain*, London: Edward Arnold.

Reitani, G. (1980) 'Politica territoriale e urbanistica in Tripolitania, 1920–40', in A. Moini (ed.) *Urbanistica Fascista*, Milan: Cinta.

Relph, E. (1987) *The Modern Urban Landscape*, London: Croom Helm.

Rex, J. (1973) *Race, Colonialism and the City*, London: Routledge.

Rex, J. (1981) 'A working paradigm for race relations research', *Ethnic and Racial Studies* IV (1): 1–24.

Rex, J. (1982) 'Convergences in the sociology of race relations and minority groups', in T. Bottomore *et al.* (eds) *Sociology. The State of the Art*, London: Sage, pp. 173–200.

Ribeiro, A. (1989) 'Capital and real estate development in Buenos Aires, 1880–1930', *International Journal of Urban and Regional Research*, 13, 1.

Richards, J.R. (ed.) (1961) *New Buildings in the Commonwealth*, London: Architectural Press.

Roberts, B. (1978) *Cities of Peasants*, London: Arnold.

Robertson, R. (1987) 'Globalization theory and civilizational analysis', *Comparative Civilizations Review* 17: 20–30.

Robertson, R. (1988) 'The sociological significance of culture: some general considerations', *Theory, Culture and Society* 5: 3–23.

Robertson, R. and Lechner, F. (1985) 'Modernization, globalization and the problem of culture in world-systems theory', *Theory, Culture and Society* 2 (3): 103–18.

Robson, B. (1973) *Urban Growth. An Approach*, London: Methuen.

Rogers, R. (1987) 'The built form and cultural research', *Planning History Bulletin*, 9 (2): 8-12.

Ross, R. and Telkamp, G. (eds) (1985) *Colonial Cities*, Boston, Lancaster, Dordrecht: Martinus Nijhoff.

Rostow, W.W. (1960) *Stages of Economic Growth*, Cambridge: Cambridge University Press.

Runnymede Trust and Radical Statistics Group (1980) *Britain's Black Population*, London: Runnymede Trust.

Said, E. (1978) *Orientalism*, New York: Pantheon.

Salinas, P.W. (1983) 'Mode of production and spatial organization in Peru', in F.G. Moulaert and P.W. Salinas (eds) *Regional Analysis and the New International Division of Labour*, London: Kluwer-Nijhoff, pp. 79-96.

Salinas, P.W. and Moulaert, F. (1983) 'Regional political economy: an introduction and overview', in F. Moulaert and P.W. Salinas (eds) *Regional Analysis and the New International Division of Labour*, The Hague and London: Kluwer-Nijhoff Publishing, pp. 3-12.

Samson, G.C. (1910) *Houses, Villas, Cottages and Bungalows for Britishers and Americans Abroad*, London: Crosby Lockwood.

Sandercock, L. (1975) *Cities for Sale*, Melbourne: Melbourne University Press.

Santos, M. (1975) *The Shared Space. The Two Circuits of Urban Economy in the Underdeveloped Countries and their Spatial Repercussions*, London: Methuen.

Saueressig-Schreunder, Y. (1986) 'The impact of British colonial rule on the urban hierarchy of Burma', *Review* 10 (2): 245-77.

Saunders, P. (1981) *Social Theory and the Urban Question*, London: Hutchinson (2nd revised edition, 1984).

Scargill, D.I. (1979) *The Form of Cities*, New York: St Martin's Press.

Schneider, J. (1986) 'The development of resorts, Lima, Peru', term paper, Housing Studies programme, Graduate School, Architectural Association, London.

Schwerdtfeger, F. (1982) *Traditional Housing in African Cities*, Chichester: John Wiley.

Scull, A. (1982) 'A convenient place for inconvenient people: the Victorian lunatic asylum', in A.D. King (ed.) *Buildings and Society: Essays on the Social Development of the Built Environment*, London: Routledge & Kegan Paul.

Shapiro, S.G. (1973) 'Planning Jerusalem: the first generation, 1917-1968', in D.H.K. Amiran *et al.* (eds) *Urban Geography of Jerusalem*, Jerusalem: Massada Press, pp. 139-53.

Simmons, C. and Kirk, R. (1981) 'Lancashire and the equipping of the Indian cotton mills: a study of textile machinery supply, 1854-1939', proceedings of the Seventh European Conference on Modern South Asian Studies, School of Oriental and Asian Studies, London, July.

Simon, D. (1986) 'Desegregation in Namibia: the demise of urban apartheid?', *Geoforum* 17 (2) 289-307.

Simon, D. (1989) 'Colonial cities, post colonial Africa and the world-economy: some pertinent issues and questions', *International Journal of Urban and Regional Research* 13 (1).

Slater, D. (1980) 'Towards a political economy of urbanization in peripheral capitalist societies. Problems of theory and method with illustrations from Latin America', *International Journal of Urban and Regional Research* 4 (3): 27–51.

Slater, D. (1986) 'Capitalism and urbanization at the periphery', in D. Drakakis-Smith (ed.) *Urbanization in the Developing World*, London: Croom Helm.

Smith, M.P. (1965) *The Plural Society in the British West Indies*, Oxford: Oxford University Press.

Smith, M.G. (ed.) (1983) 'Structuralist urban theory: a symposium', *Comparative Urban Research* 9: 5–70.

Smith, M.P. (1980) *The City and Social Theory*, Oxford: Oxford University Press.

Smith, M.P. (ed.) (1984) *Cities in Transition*, London: Sage.

Smith, M.P. and Feagin, J.R. (1987) *The Capitalist City*, New York: Blackwell.

Smith, T.R. (1869) 'On buildings for European occupation in tropical climates, especially India', *Proceedings of the RIBA*, 1868–9, 1st series, 18: 197–208.

Soja, E., Morales, R. and Wolff, G. (1983) 'Urban restructuring: an analysis of social and spatial change in Los Angeles', *Economic Geography*, 59: 195–230.

Soja, E.W. and Weaver, C.E. (1976) 'Urbanization and underdevelopment in East Africa', in B.J.L. Berry (ed.) *Urbanization and Counter-Urbanization*, Beverly Hills, CA and London: Sage, pp. 233–66.

Sovani, N.V. (1966) 'The analysis of overurbanization' in (ibid) *Urbanization and Urban India*, Bombay, London, New York: Asia Publishing House, pp. 1–13.

Spencer, J.E. and Thomas, W.L. (1948) 'The hill stations and summer resorts of the orient', *Geographical Review* 39 (4): 637–51.

Spodek, H. (1980) 'Studying the history of urbanisation in India', *Journal of Urban History* 6: 251–95.

Steinberg, F. (1984) 'Town planning and the neocolonial modernization of Colombo', *International Journal of Urban and Regional Research*, 8 (4) 530–48.

Stevens, P.H.M. (1955) 'Planning legislation in the colonies', *Town and Country Planning*, March.

Sunkel, O. and Fuenzalida, E.F. (1979) 'Transnationalisation and its national consequences', in J.J. Villamil (ed.) *Transnational Capitalism and National Development*, Sussex: Harvester Press, pp. 67–93.

Sutcliffe, A. (1981) *Towards the Planned City: Germany, Britain, the United States and France, 1790–1914*, Oxford: Blackwell.

Swanson, M.W. (1969) 'Urban origins of separate development', *Race* 10: 31–40.

Swanson, M.W. (1970) 'Reflections on the urban history of South Africa', in H.L. Watts (ed.) *Focus on Cities*, Durban: Institute of Social Research, University of Natal.

Swinton, G. (1912) 'Planning an imperial capital', *Garden Cities and Town Planning* 2 (4) (NS).

Tabb, W. and Sawers, L. (eds) (1978) *Marxism and the Metropolis*, New York: Oxford University Press (2nd edn, 1984).

Tarapor, M. (1984) 'Art education in Imperial India. The Indian schools of art' in K. Ballhatchet (ed.) *Changing South Asia: City and Culture*, Hong Kong: Asian Research Service, pp. 91–8.

Tarn, J. (1971) *Working Class Housing in Nineteenth-Century Britain*, London: Lund Humphries.

Tarn, J.N. (1974) *Five Per Cent Philanthropy*, Cambridge: Cambridge University Press.

Taylor, M.J. and Thrift, N.J. (eds) (1982) *The Geography of Multinationals*, London: Croom Helm.

Taylor, P.J. (1985) *Political Geography. World-Economy, Nation-State and Locality*, London: Longman.

Telkamp, G.J. (1978) 'Urban history and European expansion. A review of recent literature concerning colonial cities and a preliminary bibliography', *Intercontinenta 1*, Leiden: Centre for the History of European Expansion, University of Leiden.

Thane, P. (1986) 'Financiers and the British state: the case of Sir Ernest Cassel', *Business History*, 28 (1): 80–99.

Thomas, B. (1972) *Migration and Urban Development. A Reappraisal of British and American Long Cycles*, London: Methuen.

Thompson, F.M.L. (ed.) (1983) *The Rise of Suburbia*, Leicester: Leicester University Press.

Thomson, J.K.J. (1982) 'British industrialization and the external world: a unique experience or an archetypal model', in M. Bienefeld and M. Godfrey (eds) *The Struggle for Development. National Strategies in an International Context*, Chichester: Wiley, pp. 65–92.

Thrift, N.J. (1986) 'The geography of international economic disorder', in R.J. Johnston and P.J. Taylor (eds) *A World in Crisis? Geographical Perspectives*, Oxford: Blackwell, pp. 12–67.

Thrift, N.J. (1987a) 'The fixers: the urban geography of international commercial capital', in J. Henderson and M. Castells (eds) *Global Restructuring and Territorial Development*, London: Sage.

Thrift, N.J. (1987b) 'Serious money. Capitalism, class, consumption and culture in late twentieth century Britain', paper presented to the IBG Conference on 'New Directions in Cultural Geography', London, September.

Thrift, N., Leyshon, A., and Daniels, P. (1987), 'Sexy greedy. The new international financial system, the City of London and the South East of England', paper for the Sixth Urban Change and Conflict Conference, University of Kent, September (forthcoming with previous papers in N. Thrift and A.L. Leyshon *Making Money. The City of London and Social Power in Britain*, London: Routledge).

171

Timberlake, M. (ed.) (1985) *Urbanization in the World-Economy*, London: Academic Press.

Timberlake, M. (1987) 'World-system theory and the study of comparative urbanization', in M.P. Smith and J.R. Feagin (eds) *The Capitalist City*, Oxford: Blackwell, pp. 37–65.

The Times (1984a) 'Company to cash in on leisure', 10 January, p. 3.

The Times (1984b)' Supervised bungalows in big demand', 16 January, p. 3.

Tomlinson, H. (1980) 'Design and reform: the "separate system" in the nineteenth century English prison', in A.D. King (ed.) *Buildings and Society: Essays on the Social Development of the Built Environment*, London: Routledge & Kegan Paul, pp. 94–122.

United Nations (1971) *Climate and House Design*, New York: United Nations.

Urry, J. (1987) 'On the waterfront', *New Society*, 14 August, pp. 17–19.

Vance, J.E. (1971) 'Land assignment in the pre-capitalist, capitalist and post-capitalist city', *Economic Geography* 47: 101–20.

Van Oss, A. (1980) 'The colonial city in Spanish America', CEDLA, University of Amsterdam (unpublished).

Vining, D.R. Jr. and Kontuly, T. (1978) 'Population dispersal from major metropolitan regions: an international comparison', *International Regional Science Review* 3: 49–73.

Walker, R.A. (1978) 'The transformation of urban structure in the nineteenth century and the beginnings of suburbanization', in K. Cox (ed.) *Urbanization and Conflict in Market Societies*, Chicago, IL: Maaroufa Press, pp. 165–212.

Walker, R.A. (1981) 'A theory of suburbanisation; capitalism and the construction of urban space in the United States', in M. Dear and A.J. Scott (eds) *Urbanization and Urban Planning in Capitalist Society*, London: Methuen, pp. 383–430.

Wallerstein, I. (1974) *The Modern World System*, London: Academic Press.

Wallerstein, I. (1976) 'The three stages of African involvement in the world-economy', in P.C.W. Gutkind and I. Wallerstein (eds) *The Political Economy of Contemporary Africa*, London and Beverly Hills: Sage, pp. 30–52.

Wallerstein, I. (1979) *The Capitalist World-Economy*, Cambridge: Cambridge University Press.

Wallerstein, I. (1984) *The Politics of the World-Economy*, Cambridge: Cambridge University Press.

Wallerstein, I. (1987) 'World-systems analysis', in A. Giddens and J.H. Turner (eds) *Social Theory Today*, Cambridge: Polity Press.

Walton, J. (1976) 'Political economy of world urban systems: directions for comparative research', in J. Walton and L. Massotti (eds) *The City in Comparative Perspective*, London: Sage, pp. 301–13.

Walton, J. (1979) 'Urban political economy: a new paradigm', *Comparative Urban Research* 7 (1): 5–17.

Walton, J. (1984) 'Culture and economy in shaping urban life: general issues and Latin American examples', in J.A. Agnew, J. Mercer, and

D. Sopher (eds) *The City in Cultural Context*, New York and London: Allen & Unwin, pp. 76–93

Walton, J. (1985) *Capital and Labor in the Urbanized World*, London, New York: Sage.

Western, J. (1985) 'Undoing the colonial city', *Geographical Review*, 73 (3): 335–57.

Wheatley, P. (1969) *City as Symbol*, London: University College.

White, L.W.T. *et al.* (1948) *Nairobi: Master Plan for a Colonial Capital*, London: HMSO.

Williams, E. (1944) *Capitalism and Slavery*, Chapel Hill: University of North Carolina Press.

Williams, R. (1973) *The Country and the City*, Oxford: Oxford University Press.

Williams, R. (1976) *Keywords. A Vocabulary of Culture and Society*, London: Fontana.

Winter, R. (1980) *The California Bungalow*, Los Angeles, CA: Hennesey & Ingalls.

Wolf, E. (1982) *Europe and People without History*, Berkeley: University of California Press.

Woodruff, W. (1979) 'The emergence of an international economy, 1700–1914', in C.M. Cipolla (ed.) *The Fontana Economic History of Europe*, The Emergence of Industrial Societies, 2, London: Collins pp. 656–737.

Wright, G. (1987) 'Tradition in the service of modernity: architecture and urbanism in French colonial policy, 1900–1930', *Journal of Modern History* 59: 291–316.

Wright, G. (1991) *At Home and Abroad: French Colonial Urbanism, 1880–1930*, Univeristy of Chicago (forthcoming).

Wright, G. and Rabinow, P. (1981) 'Savoir et pouvoir dans l'urbanisme moderne colonial d'Ernest Hebrard', *Cahiers de la Recherche Architecture*.

Zukin, S. (1980) 'A decade of the new urban sociology', *Theory and Society* 9: 575–601.

An earlier version of Chapter 2 was published in Ross and Telkamp (eds) *Colonial Cities. Essays on Urbanism in a Colonial Context*, Martinus Nijhoff for the Leiden University Press, Dordrecht, Boston, and Lancaster, 1985; some sections of Chapter 3 first appeared in a paper in 'Exporting planning: the colonial and neo-colonial experience', *Urbanism Past and Present*, 5 (1977–8) and in a revised form in G. Cherry (ed.) *Shaping an Urban World*, Mansell, London, 1980; an earlier version of Chapter 4 appeared in *Urban History Yearbook, 1983*, Leicester University Press, and of Chapter 5, in a festschrift edition of *Habitat International*, vol. 7, no. 5/6 1983 (Pergamon Press) for Professor Otto Koenigsberger; an earlier version of Chapter 6 appeared in *Environment and Planning D: Society and Space*, 1984, 2 (Pion). Whilst the papers have been considerably revised since first published, I am grateful to the publishers for their permissions.

INDEX

Medina 63
Meegan, R. 130
Meknes, 31, 153
Melbourne 140, 142
Metcalfe, T.R. 60, 145
Mexico 22, 69, 154
Mexico City 22, 69
'miasmic theory' 41
Middle East 4, 80, 83, 99, 103, 138; colonial city 38, 42; planning 153
migration 127-8, 143; from core to periphery 6, 25; from periphery to core 68-70, 76, 82, 152; rural-urban 5, 30, 50-2, 56-7
military settlements and engineers 38, 50, 61-2, 145
Mingione, E. 71
mining/minerals 5, 14, 30, 49, 120, 144, 146, 154
Misra, R.P. 83, 95
Mitchell, J.C. 35, 57
modern building form 83-4, 94, 95
modern planning 8, 9, 47-8, 61-3
modern societies see independence; post-colonialism
modern towns (new towns) 9, 62, 65
Mombasa 140, 141, 146
Montreal 140, 142, 147
Morocco 9, 46, 153, 154; colonial city 22, 31, 41-3; independence 151; planning 46, 62-3
Morris, J. 148
motivation for colonization 28-31, 36; see also capitalism; resources
Moulaert, F. 134-5
Mozambique 62
Mullins, P. 60, 74-5, 149
multinational corporations (transnationals) 50, 53, 69, 76-7, 79, 81, 110, 153
multinational culture and capital see world-economy
Murphy, R. 153
Mussoorie 150
Muthesius, S. 128

Nairobi 22, 25, 42, 151; planning 46, 52, 56
Naqvi, H.K. 153
Nassau 140, 142
Natal 132, 141, 146
'nativism' 26
Neild, S.M. 36, 109

neocolonialism see independence; modern; post-colonialism
Netherlands 3, 68, 98, 138; building form 101, 106, 108, 127; colonial city 16, 25-6, 28-9, 32, 43; planning 45, 46
New Delhi see Delhi
new urban studies 102-3
New York 17, 22, 66, 69
New Zealand 127, 140, 142, 148
Newcastle 139, 152
Niger 78, 151
Nigeria 46; building and international division of labour 140, 141, 147, 153-4; colonial city 21, 23; independence 151; planning 48, 51, 55-6, 60, 64
Nikosia 140, 141
Nilsson, S. 58
North Africa 9, 95, 153-4; building 109, 140; colonial city 14, 22, 30-1, 34, 37-9, 40-3; independence 151; planning 46, 61-3; tenure 42-3; urban history and world-system 72, 78
North America: building form 103, 105-6, 108, 113-14, 124, 126; building and international division of labour 138, 140, 142, 147-9; colonial city 17, 21, 22, 25, 38, 40; colonialism 3-4, 6; culture and political economy 95, 97; planning 45, 46, 50, 62; urban history and world-system 69, 70, 73, 75, 81; see also Canada; United States
Norway 127
Norwich 139, 153
Nottingham 152
Nyasaland 44, 46, 151

oil prices 65
Oldenburg, V.T. 40, 60, 110
Oliver, P. 55, 66
O'Neill, S. 81
Oosterhoff, J.L. 29
OPEC 65
order see power
organization of colonial city 27-8, 31-3; see also institutions; legislation; planning; segregation
Ottawa 142
Ottoman Empire 3
ownership 124-6